Night Comes to the Cretaceous

NIGHT COMES TO THE CRETACEOUS

DINOSAUR EXTINCTION AND THE TRANSFORMATION OF MODERN GEOLOGY

JAMES LAWRENCE POWELL

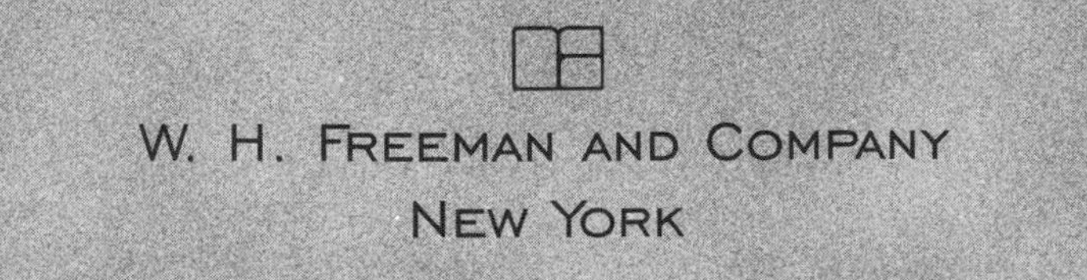

W. H. FREEMAN AND COMPANY
NEW YORK

To My Children: Marla, Dirk, and Joanna
And to Gene Shoemaker: The Pathfinder

Cover illustration and design: Roy Wiemann
Text design: Diana Blume

Cover photos: Top, *Mamenchisaurus,* detail from "Crossing the Flats" © 1986 by Mark Hallett, Salem, Oregon; bottom left, dinosaur footprints, New York Public Library Picture Collection; bottom right, the Cretaceous/Tertiary boundary interval, Raton Basin, Colorado and New Mexico, photo by Glenn Izett.

Library of Congress Cataloging-in-Publication Data

Powell, James Lawrence, 1936–
Night comes to the Cretaceous : dinosaur extinction and the transformation of modern geology / James L. Powell
p. cm.
Includes bibliographical references and index.
ISBN 0-7167-3117-7
1. Catastrophes (Geology) 2. Extinction (Biology) 3. Dinosaurs. 4. Title.
QE506.P735 1998
576.8'4—dc21 98–13192
CIP

Printed in the United States of America

Second printing, 1998

CONTENTS

PREFACE

What killed the dinosaurs? For 150 years, that question has stumped even the best scientists. But no longer. At last the great mystery has been solved. The story of the solution is fascinating in its own right, and the answer helps us to understand our place in the universe: It raises the revolutionary possibility that the history of the earth, and of life upon it, has been altered repeatedly by a nearly invisible, and previously unrecognized, cosmic process. It forces us to ponder the role of chance in the solar system, even in the evolution of our species. And by helping geologists to cast off outmoded dogma and to acknowledge that our planet is subject to the same processes as other bodies in the solar system, it has transformed the science.

My interest in the story of dinosaur extinction developed over many years as a geologist, professor, and museum director. It peaked one day in December 1993, when the latest issue of *Nature* landed on my desk.[1] Staring up at me from the cover was a photograph of a zircon crystal, scarred with the crisscrossing fractures that I knew indicated extreme shock pressure. "Fingerprinting impact debris," the caption read. The accompanying article revealed how analysis of the shattered zircon helped to corroborate the radical theory that scientists from the University of California at Berkeley had proposed in 1980 to explain the disappearance of the dinosaurs. Fascinated by the detective work involved, I conceived of writing a book that would tell the story of how the great mystery was solved, and explore the extraordinary implications of the new theory.

The story turned out to be richer than I had ever imagined. It provides the best lesson in this century of how scientists challenge and overthrow orthodoxy, and how science *really* works, not in the mythical ivory tower but down in the trenches. It shows how the suggestion that the dinosaurs died in a catastrophe brought on by the collision of an extraterrestrial object violated geologic dogma not once but twice, and incurred the wrath of many. The cast of characters who took part in the bitter debates that followed shows all too clearly that scientists are passionate, and sometimes flawed, human beings.

My exploration of the claim for an extraterrestrial cause drew on many different fields of science: vertebrate paleontology, micropaleontology, evolutionary biology, rare-metal chemistry, astronomy, magnetism, statistics, geologic age dating, and the physics of nuclear explosions. Reviewing the evidence from these many disciplines and the writings of geologist colleagues gave me a new appreciation of just how much my field has changed.

Natural history, and geology in particular, has always fascinated me. The founders of my hometown, the small college community of Berea, Kentucky, wisely decided to site it "where the mountains kiss the bluegrass," at the foot of the Cumberland Mountains. To a boy interested in nature, the setting was perfect. Heading out of town on my bicycle, I soon found myself among an abundance of birds, rocks, fossils, butterflies, and Native American artifacts. Like all children, I was fascinated by dinosaurs, which seemed more like creatures of fantasy than ones that had actually lived.

Having been raised in a small academic community, it was perhaps also natural that I would earn a doctorate in my chosen field, geology, and then go on to another college town, Oberlin, Ohio, to teach. One thing led to another and eventually I found myself a college president and then a museum director. My move to the Natural History Museum of Los Angeles County has been especially gratifying because here I am surrounded by birds, rocks, fossils, butterflies, and artifacts, like the ones that got me started so many years ago. But now I preside over 35 million of them!

Many colleagues have helped me in writing this book, though I alone am responsible for any errors of fact or interpretation. The following read the manuscript and made helpful suggestions: Alfred Fischer, Peter Griffin, John Harris, Brian Huber, Adriana Ocampo, Kevin Pope, Donald Reich, Robin Simpson, and Peter Ward. My assistants Pat Reynolds and the late Patricia Barron aided me in many ways, as did Museum Librarians Donald McNamee and Mark Herbert. Former W. H. Freeman and Company Editor Elizabeth Knoll, now at Harvard University Press, had faith that a book on the Alvarez theory would serve a useful purpose. Jonathan Cobb, formerly Senior Editor at W. H. Freeman, was a joy to work with and contributed significantly to the quality of the book. My wonderful agent, Barrie Van Dyck, never gave up. Finally, my wife, Joan, and our daughter, Joanna, were patient beyond any reasonable expectation during the lengthy process of writing this book while I was holding down a full-time job directing a museum. To all of them I am grateful.

Prologue

The Greatest Mystery

The dinosaurs were among the most successful creatures ever to live. They reigned for 160 million years, far longer than the few million that our genus, *Homo*, has so far existed. Dinosaurs came in all sizes and shapes: small and large, fast and slow. Except for today's blue whale, no creature has been larger, and surely none have been as frightening. Some, like *Apatosaurus*, the 80-foot, 30-ton thunder lizard whose tread shook the ground like an earthquake, were herbivores. Others, like the 45-foot-long, 20-foot-tall *Tyrannosaurus rex*, a favorite of film and of children, were carnivores.

But after having thrived for millions of years, suddenly, in a relative instant of geologic time, the giant reptiles vanished completely and forever, leaving in their wake the greatest scientific mystery: What killed the dinosaurs? The mystery deepens when we learn that it was not only the dinosaurs that became extinct 65 million years ago, but 70 percent of all species on the earth. The geologic boundary that marks the time of their demise is used to define the end of one great geologic era, the Mesozoic (middle life), and the beginning of the one in which we live, the Cenozoic (modern life). Curiously, the snakes, turtles, and crocodiles, which one would have thought were enough like the dinosaurs to have met a similar fate, survived. What could kill every single dinosaur but spare these other reptiles?

Since the time 150 years ago when the dinosaurs were first discovered, hundreds of scientists have struggled to solve this mystery. Glenn Jepsen of Princeton University, in a 1964 article, summed up the many solutions that had been proposed—some worthy, some feeble, some surely facetious:

> Authors with varying competence have suggested that dinosaurs disappeared because the climate deteriorated (became suddenly or slowly too hot or cold or dry or wet), or that the diet did (with too much food or not

> enough of such substances as fern oil; from poisons in water or plants or ingested minerals; by bankruptcy of calcium or other necessary elements). Other writers have put the blame on disease, parasites, wars, anatomical or metabolic disorders (slipped vertebral discs, malfunction or imbalance of hormone and endocrine systems, dwindling brain and consequent stupidity, heat sterilization, effects of being warm-blooded in the Mesozoic world), racial old age, evolutionary drift into senescent overspecialization, changes in the pressure or composition of the atmosphere, poison gases, volcanic dust, excessive oxygen from plants, meteorites, comets, gene pool drainage by little mammalian egg-eaters, overkill capacity by predators, fluctuation of gravitational constants, development of psychotic suicidal factors, entropy, cosmic radiation, shift of Earth's poles, floods, continental drift, extraction of the moon from the Pacific Basin, drainage of swamp and lake environments, sunspots, God's will, mountain building, raids by little green hunters in flying saucers, lack of even standing room in Noah's Ark, and paleoweltschmerz.[1]

Such a long list of theories, so many of them downright silly and contradictory, suggests a mystery so deep as to be beyond us. Many people must have shared the attitude of humorist Will Cuppy: "The age of dinosaurs ended because it had gone on long enough and it was all a mistake in the first place."

Why, given the intense interest that dinosaurs have aroused since their discovery a century and a half ago, did it take so long to solve the riddle of their extinction? One reason is that in the historical sciences—geology, archaeology, paleoanthropology—definitive answers are particularly hard to come by. Scientists in these fields do not have the advantage of being able to design and then conduct experiments, as is done in chemistry, physics, and many areas of biology. Rather, they have to operate more as detectives. The "experiments" were conducted long ago by nature; scientists working today literally must pick up the pieces and try to interpret them. A "crime"—in this case, dinosaur extinction—is discovered, sometimes by accident. In the mystery novel, the clues accumulate and, before we realize it, the clever detective has identified the culprit. But the mysteries of the earth, like real crimes, are not always so easily solved.

Paleontology, the study of ancient life, is an especially difficult science in which to arrive at definitive answers. We can learn about prehistoric animals only from their fossilized remains, and yet many had no hard parts and therefore left no trace. Of them, we can never learn anything. Other organisms were bony; but all too often, before they could be fossilized, their bones dissolved or weathered away. The fossils that did form had a way of winding up in rocks other than those in which, millions of years later, we happen to be searching. Paleoanthropologist Meave Leakey, of the famous family whose work has transformed the study of human origins, describes her nearly futile hunt for human bone in a new field area as four years of hard work producing only three nondescript scraps.[2] To take a different kind of example, in the winter after the great Yellowstone fires of 1988, thousands of elk perished from extreme cold coupled with lack of food. Late the following spring, their carcasses were strewn everywhere. Yet only a few years later, bones from the great elk kill are scarce. The odds that a single one will be preserved so that it can be found 65 million years from now approach zero. At best we can expect to find fossil evidence of only a tiny fraction of the animals that once lived. The earth's normal processes destroy or hide most of the clues.

Still, it comes as a shock to realize that in spite of the intense popular and scientific interest in the dinosaurs and the well-publicized efforts of generations of dinosaur hunters, only about 2,100 articulated dinosaur bones (two or more aligned in the same position as in life) have ever been found.[3] Conclusions about the life and death of the dinosaurs thus rest on a small sample indeed. As a result, in spite of the fascination they hold, our knowledge of dinosaurs has progressed slowly. Until recently, as Jepsen's long list shows, the door to speculation about their demise has been wide open. Almost any notion could be proposed and avoid being refuted on the evidence.

A second reason that solving the mystery of dinosaur extinction proved so difficult is that geologists, having correctly dismissed most of the reasons on Jepsen's list, were questioning only the usual suspects: changes in sea level, geography, and climate. Throughout the history of the earth continents have grown and eroded, seafloors have divided, spread, and closed, and the earth has moved closer and farther away from the sun, all causing countless changes in sea level, geography, and climate. Perhaps one of these familiar mechanisms, reaching a rare extreme, brought down the great beasts. Yet surely creatures that had lived for 160 million years had survived many such changes and others that we can only imagine. How could a decline in sea level, even an extreme one, have caused their demise? Would it not merely have opened up more land on which they could roam? These explanations seemed contrived, a bit like accusing the staid and familiar butler of actually being the maniac who murdered the family he had served faithfully for decades. Suppose instead that the culprit was a complete stranger who appeared out of nowhere, entered violently, stayed only long enough to do the deed, and then vanished as suddenly as he had appeared. And 65 million years ago at that. Now there's a mystery to defy even the best detective.

Another problem with the usual list of geologic suspects is that although most earthly processes operate slowly, for decades scientists believed that the dinosaurs had expired suddenly. How could a gradual process cause a rapid extinction? Over the last several decades some additional dinosaur fossils have been found, but none in the few centimeters just below the Mesozoic–Cenozoic boundary; instead, they have seemed to peter out farther down, in rocks older than the boundary. The absence of dinosaur bones right at the boundary led most specialists to discard the long-standing view that the dinosaurs had expired suddenly, and to become persuaded that instead they had started to wane millions of years before the end of the Mesozoic. Contrary to public opinion, the vertebrate paleontologists thought, to paraphrase Dylan Thomas, that the terrible lizards had gone gently into the night of extinction. Not only was their disappearance not mysterious, it was inevitable, for extinction is the fate of all species. After all, the average species survives for only about 4 million years; over 95 percent of those that have ever lived are extinct. Hence, dinosaur extinction seemed to need no special explanation—it was simply the way things were and are today. The great mystery had been converted into something mundane.

A pair of scientists writing in 1979 expressed it this way: The dinosaurs "may have succumbed to a series of environmental disasters, some dying of thirst, others of hunger, and the stragglers may have perished because the reduced population density rendered the community unviable."[4] What a sorry end to 160 million years of supremacy! In this view, the dinosaurs had played the game of evolution longer than most, but in the end, they too had lost, going out, in T. S. Eliot's phrase, "not with a bang but a whimper."[5]

Near the end of the 1970s, dinosaur specialist Dale Russell wrote a review article in which he selected and examined those theories that seemed scientifically plausible out of the scores that had been offered.[6] In the end, he had to conclude that only one held up to scrutiny: The dinosaurs died because an exploding star or supernova—a literal death ray—had spread lethal radiation effects throughout our region of the galaxy. Russell ended his paper on this pessimistic note: "If a fundamental deficiency were found in the supernova model . . . the disappearance of the dinosaurs would remain an outstanding mystery of the geologic record."[7]

The theories that Russell examined shared three shortcomings: First, the evidence to support them had been sifted countless times and proven inconclusive. Second, new evidence was scarce. Third, the theories made few or no testable predictions. This last weakness is crucial: To be useful, a theory must make predictions that can be tested—it must lead to questions that can be examined in the field and laboratory. A theory that suggests no further action could be correct for all we know, but since it can neither be corroborated nor falsified, there is no way to find out. Theories that cannot be tested are like rocks in a stream, around which the river of scientific progress must flow.

All the plausible theories of dinosaur extinction were based on the assumption that the earth has always worked as it does today, a reasonable supposition that geologists have employed almost from the beginning of their science. In this large-scale view, change derives not from sudden catastrophe but from deliberate and inexorable processes—erosion, deposition, the long-term rise and fall of the sea, the uplifting and downwearing of continents—whose full effects can only be seen after the passage of hundreds of thousands, even millions, of years.

But what if one were to set aside the assumption of slow change and ask whether an event as rare as dinosaur extinction might not have had an equally rare cause—a catastrophe that would appear only on the same time scale, say, every 50 million to 100 million years? If that were the case, the geologic sleuths needed to detect an event that no human being had ever seen, the evidence for which might be buried 65 million years in the past. To make the task even more difficult, the cause turned out to be one in which, at the start, almost no geologists believed.

Success at solving the mystery of dinosaur extinction required a rare concatenation of circumstances and a healthy dose of good luck. Though, as often happens in science, more than one person was on the right trail; a remarkable pair got there first. One of the pair was a Nobel Prize–winning physicist with no professional experience in geology, a problem solver unable to resist going after the biggest scientific mystery and who, late in his career, could afford to do so with impunity. Luis Alvarez was a man of unusual energy and curiosity. Without these qualities, he would have had neither the interest nor the willingness to tackle a problem so far outside his discipline. A scientist at an earlier career stage likely could not have taken the risk of working in a field so far outside his or her own, where the chances of failure or success were harder to predict.

The other person was an experienced geologist who understood the complexities of the dinosaur puzzle but, having worked in other branches of earth science, had no previously announced position on dinosaur extinction to defend. A meticulous scientist, he helped to build the case among geologists themselves that something far outside their experience had caused the extinction. His presence kept opponents from dismissing the new theory as merely the work of an arrogant, geologically challenged, physicist.

Success also depended on a third and special factor—the relationship between the two scientists. Luis and Walter Alvarez were father and son. Their kinship kept at bay the kinds of issues that so often retard progress—mistrust, rivalry, priority, arguments over who should be the senior author on a research paper. Walter's gentlemanly style helped to offset the wrath generated among geologists by his pugnacious father.

A fourth critical element was that often essential ingredient in scientific discovery: serendipity. Luis and Walter Alvarez were lucky. But although the initial discovery that eventually led to the solution of the riddle was accidental, the minds of Luis and Walter Alvarez, in combination, were well prepared to grasp the opportunity that serendipity presented. And, the timing was right. As both Jepsen's amusing list and Russell's scientific analysis demonstrated, geologists really had little to show for 150 years of trying to find out what had killed the dinosaurs. No one really knew, and therefore no strong rival theory existed. Furthermore, by the 1970s, the exploration of space, upon which so much of the solution would hinge, was in full bloom; those who had been paying attention knew that impact catastrophes were rife in the history of the solar system—just look at the craters on the moon. Finally, by 1980, geologists had developed tools to recognize the much more obscure and rare craters that exist on the earth.

As Russell was writing his article, the Alvarezes and their coworkers at Berkeley were pursuing one of the theories that Russell had mentioned in passing but for which no evidence had existed: the idea that at the end of the Mesozoic era, a large meteorite had struck the earth and thrown up such a vast cloud of dust that it darkened the sky, lowered world temperatures by many degrees, halted photosynthesis, disrupted the food chain, and thus gave rise to a great mass extinction.

Shakespeare's Caesar tells us "The fault, dear Brutus, is not in our stars, But in ourselves." The Alvarez theory held that the fault was *not* in the dinosaurs themselves but, almost literally, in their stars! They died through no deficiency of their own, but simply as a result of being in the wrong place at the wrong time. In contrast to almost all the possibilities on Jepsen's list, the Alvarez theory made predictions (Luis counted 15) and therefore it could be tested. The most striking prediction was that somewhere on the surface of the earth there may lie hidden a crater exactly 65 million years old.

The new theory ran into trouble immediately. To accept that the Alvarezes were right, the vertebrate paleontologists would have had to admit that they were wrong: The dinosaurs had gone extinct suddenly after all. Even more seriously, the theory required all geologists to accept that one of the greatest events in earth history was explained by a random catastrophe. The trouble was, the notion of catastrophism as the cause of geologic events had been cast off 150 years previously; in fact, abandoning it had been central to the birth of geology as a modern science. In the decades since, geologists had been eminently successful in explaining earth history by appealing only to slow, noncatastrophic processes. To fall back now on a catastrophe to explain dinosaur extinction would seem to be a return to the dark ages of their science.

A group of scientists led by Professor Charles Officer, now retired from Dartmouth College, not only believed that the Alvarez theory was wrong, as did many, but actively set out to refute it. They published hundreds of papers presenting evidence that they believed contradicted the theory. But in Luis Alvarez, the critics found a brilliant scientific opponent who loved a good fight. The debate descended to one of the all-time lows of scientific discourse. Insults were thrown with abandon and careers were damaged. A Berkeley paleontologist labeled the theory "codswallop;" Luis responded that paleontologists were nothing more than "stamp collectors"—to a physicist, the ultimate insult. Critics reached back to 1954 to dredge up Luis's controversial role in the notorious hearings that led to the dismissal of J. Robert Oppenheimer, father of the atom bomb, as a security risk. The debate turned ugly indeed, with ample blame to go around. The making of science is often not the pretty sight that textbooks and scientific papers written after the fact would have us believe.

Geologists also resisted because, in another area of their discipline, they had just weathered a scientific revolution. The late Thomas Kuhn, and other students of the history of science, have shown how almost all scientists work within a common set of understandings of their discipline—within the confines of a particular ruling paradigm.[8] Sooner or later, new evidence is found that the current paradigm cannot explain, and the paradigm needs to be replaced. The more senior scientists tend to cling to the fading paradigm, on which they have built their careers and reputations, some going to their graves obstinate in their belief. The young turks, and others who are more flexible or iconoclastic, rally to exploit the new paradigm. But all good things must come to an end, and eventually the former young turks, now scientific senior citizens, are themselves displaced by a new group of scientific provocateurs. All of this has great human cost, as scientists are no more inclined than anyone else to admit they were wrong.

In the 1960s and 1970s, geologists had made a revolutionary paradigm shift: They had finally come to accept a modern version of the theory that the continents are not fixed on the surface of the earth, but drift, even colliding with each other to throw up mountain ranges such as the Appalachians and the Urals. The theory of continental drift, originally proposed in 1915, had been largely forgotten by mid-century. Then, suddenly, a new kind of evidence became available from studying the magnetism of the ocean floors. It showed that continents do not actually drift through the earth's mantle (the second zone of the layer-cake structure of crust, mantle, and core); rather, the surface of the earth is divided into giant, moving plates that carry continents along on them, as though they were riding piggyback on a huge, ponderous conveyor belt. Now, with only a decade to adapt to this revolution, geologists were being asked (and by a physicist!) to accept another: that Mother Earth does not always evolve as a result of imperceptible change conducted over deep geologic time, but sometimes, perhaps even often, as the result of sudden and apparently random catastrophes.

The meteorite impact theory proved even more difficult for geologists to swallow than continental drift, for it appeals to a deus ex machina, exactly the opposite of the way they normally work and think. Impact theory asks geologists to look, not down at the familiar terra firma that has always drawn their gaze, but up at the sky—for them, an unfamiliar and uncomfortable posture. It requires that they abandon a strict interpretation of uniformitarianism—the view that all past changes can be understood by studying only processes that can be seen going on today—for well over a century the guiding paradigm of geology. But the impact theory has implications not only for geologists. It requires that biologists consider the possibility that evolution may be driven by survival not of the fittest but of the luckiest. And it requires all of us to take more seriously the role of chance in the history of our planet, our species, and ourselves.

As this book was being written, Eugene Shoemaker, the scientist whose lifetime of work helped to create and to validate this second revolution, tragically died while mapping his beloved impact craters in the Australian outback. Years of attempting to persuade his fellow scientists of the importance of meteorite impact in earth history had led Gene to observe that "most geologists just don't like to think of stones as big as hills or small mountains falling out of the sky."[9] Physicists, however, have had no such compunction.

Part I

Bolt from the Blue

Chapter 1

The Alvarez Discovery

Chance favors only the prepared mind.[1]
Louis Pasteur

The Father

In a picture taken at the MIT Radiation Laboratory during World War II, his hat at a jaunty angle and a cigarette dangling from his lips, a cocky smile on his face and a coil of wire strung around his neck, Luis Alvarez appears not as the stereotypical dull, introverted scientist but more like a cross between Indiana Jones and Humphrey Bogart (Figure 1). He was a man who lived life to the fullest and continued seeking new challenges long after most would have begun to rest on their laurels.

Luis's father, Walter Alvarez, after whom Luis named his son, was a well-known San Francisco physician who encouraged Luis's early interest in science. In 1922, when only 11, Luis surprised family friends by demonstrating the first crystal set any of them had seen. His talent for constructing apparatus of various sorts would last a lifetime. After his father moved to the Mayo Clinic in Minnesota, Luis spent his summers during high school helping out in the clinic machine shop, where he became a self-described "good pupil." He began his undergraduate years at the University of Chicago first as a chemistry major and then switched to physics, which became his constant and lifelong love. To call a person a physicist was to Luis Alvarez the highest praise. Physicists were a breed apart, cleverly applying their superior minds to the most interesting and important problems. He logically enough titled his autobiography *Adventures of a Physicist.*[2]

Luis launched his scientific career by moving to the University of California at Berkeley, where in 1936 he went to work with

FIGURE 1 Luis Alvarez at the MIT Radiation Laboratory, September 1943. [Photo courtesy of University of California Lawrence Berkeley National Laboratory.]

Ernest Lawrence, inventor of the cyclotron and one of the most influential American scientists in history. Lawrence was mentor to many who later made their mark; several, like Luis, were to follow him in winning the Nobel Prize.

"If politics is the art of the possible, research is surely the art of the soluble," wrote Sir Peter Medawar.[3] To choose a problem that cannot be solved, or one that need not be solved in order for a field to advance, is to delay the progress of both science and career. Luis Alvarez understood Medawar's point and proved a master at selecting the next important problem and designing just the piece of apparatus to solve it.

By his own description, his style was to "flit" from research problem to research problem, often to the consternation of his coworkers and students and especially his mentor, Lawrence. But during a period when physics was advancing rapidly and opportunities abounded, his method made him unusually versatile and productive. In the early days of World War II, he worked to improve the radar system that played such a crucial role in the Allied victory. From there, he went to Los Alamos to become one of the key scientists in the development of the atomic bomb. Two unusual projects that came later in his career showed his willingness to step outside physics if a problem piqued his curiosity or appeared sufficiently important. One was his use of cosmic rays, in X-ray-like fashion, to determine whether, as Luis suspected but most Egyptologists doubted, the Second Pyramid of Chephren contained undiscovered burial chambers. Luis was never reluctant to fly in the face of conventional wisdom in a field outside his own and to conduct an experiment to see who was right. In this case, however, as he readily admitted, he was proven wrong. When people who knew of his work would say, "I hear you did not find a chamber," Luis would reply, "No, we found there wasn't any chamber."[4] To seek but not to find a chamber is to find an absence of evidence. To determine that there was no chamber was to find evidence of absence. This is a distinction with a difference, the importance of which would turn up years later in the dinosaur extinction controversy.

His greatest public notice came from his investigation into the assassination of John F. Kennedy, particularly from his meticulous and inventive analysis of the Zapruder film. One frame showed JFK's head moving sharply backward as the third and fatal bullet struck, providing evidence to assassination buffs that a second gunman had fired from the front. Surely, common sense tells us, a head snapped backward by the impact of a bullet identifies the shot as having come from the front. Since by then Oswald was located behind the presidential limousine, this could only mean that a second gunman fired and therefore that there had been a conspiracy. But here common sense leads us astray: Luis showed that the laws of physics, when all (including the most gruesome) factors are taken into account, are entirely consistent with a shot from the rear causing the backward snap of JFK's head. Luis conducted experiments to prove the point but admitted that they failed to convince the buffs; years later, he would have a similar difficulty in convincing paleontologists that a random catastrophe had extinguished their dinosaurs.

The most dramatic moment of an unusually exciting early career came aboard the *Great Artiste,* the plane that accompanied the *Enola Gay* on its fateful mission over Hiroshima in August 1945. Weeks earlier Alvarez had been high above the New Mexico desert observing the Trinity atomic bomb test. He was thus one of only a handful to witness both of the first two atomic explosions. As the *Great Artiste* returned to its base on Tinian, Hiroshima destroyed in its wake, Luis wrote a letter for his son Walter, then 4 years old, to read when he was older.

> Today, the lead plane of our little formation dropped a single bomb which probably exploded with the force of 15,000 tons of high explosive. That means that the days of large bombing raids, with several hundred planes, are finished. A single plane disguised as a friendly transport can now wipe out a city. That means to me that nations will have to get along together in a friendly fashion, or suffer the consequences of sudden sneak attacks which can cripple them overnight.
>
> What regrets I have about being a party to killing and maiming thousands of Japanese civilians this morning are tempered with the hope that this terrible weapon we have created may bring the countries of the world together and prevent further wars. Alfred Nobel thought that his invention of high explosives would have that effect, by making wars too terrible, but unfortunately it had just the opposite reaction. Our new destructive force is so many thousands of times worse that it may realize Nobel's dream.[5]

While Luis's desire that the existence of nuclear weapons would prevent further wars was not fulfilled, knowledge of their destructive power may have averted a third world war, validating his hope. What no one could have predicted is that, 35 years after his letter was written, father and son would discover evidence of an explosion so enormous as to dwarf even the awful one that Luis had just witnessed.

When Dale Russell's paper on dinosaur extinction appeared in 1979, Luis Alvarez was 68 years old. Eleven years earlier, he had reached the pinnacle of scientific success, receiving the Nobel Prize in physics for his work in developing the hydrogen bubble chamber, which led to the discovery of many new subatomic particles. His Nobel citation was one of the longest on record. Ninety-nine of one hundred scientists, having had such a career, would have been content to rest on their laurels, perhaps becoming scientific elder statesmen who tread the corridors of power in Washington, D.C. Indeed, such a course might have tempted Luis, for particle physics was no longer the game that he had helped to invent. Finding new particles required higher and higher energies, and more and more money, changing particle physics into the biggest of big science. Papers emanating from such facilities as the Stanford Linear Accelerator Center and CERN, the European consortium housed in Geneva, often had dozens, even scores, of "co-authors." This was not the way Luis had succeeded in science. If not quite a lone wolf, he had at least been the alpha of a small pack.

But in the twilight of a career, even the most inventive of minds may require a spark. Although Luis confessed a notable lack of excitement about geology, his son's chosen field, in best fatherly fashion he spent many hours in discussion with Walter, each describing his scientific work to the other. Luis noted that "the close personal relationship Walt and I enjoyed dissolved the cross-disciplinary barriers."[6] One day in the mid-1970s, returning from a field trip to Italy, Walter produced a geological specimen that, for once, his father did find exciting; Luis would later say that it had "rejuvenated" his scientific career.

THE SON IN ITALY

Given his family history, it is not surprising that young Walter became a scientist himself. No one could have followed in his father's footsteps, and wisely in retrospect, Walter chose not to try but to follow his own love, geology. He earned his doctorate at Princeton under Professor Harry Hess and, until Qaddafi expelled the Americans, worked as a petroleum geologist in Libya. In 1971 he joined the faculty at the Lamont-Doherty Geological Observatory at Columbia University, where much of the research that led to the plate tectonic revolution had been done. After a few years there, Walter accepted a position at Berkeley, the university where his father was in residence and where Walter soon received tenure and the title of full professor. Had Walter remained in Libya or at Lamont-Doherty, we might still be scratching our heads over the mystery of dinosaur extinction.

During the 1970s, Walter summered in the pleasant northern Italian town of Gubbio, where he studied an unusually complete section of sedimentary rock that spanned the time from the middle of the Mesozoic era well up into the Cenozoic era. Geologists divide the Mesozoic into three periods: The oldest, the Triassic, is overlain by the Jurassic, which in turn is overlain by the Cretaceous (Figure 2). The earliest, lowest period in the Cenozoic is called the Tertiary. Its name hearkens back to an earlier time when there were thought to be four ages of rocks: primary, secondary, tertiary, and quaternary; only the last two are used today. The Mesozoic–Cenozoic boundary marks the point in geologic time at which the dinosaurs perished. It has become customary, however, to nickname the boundary for the two adjacent periods, the Cretaceous and Tertiary, rather than their eras. Geologists call it the "K–T" boundary. ("K" is used instead of "C" to avoid confusion with the older, Cambrian period; Cretaceous, which comes from the Latin *creta*, for chalk, also happens to be *Kreide* in German.)

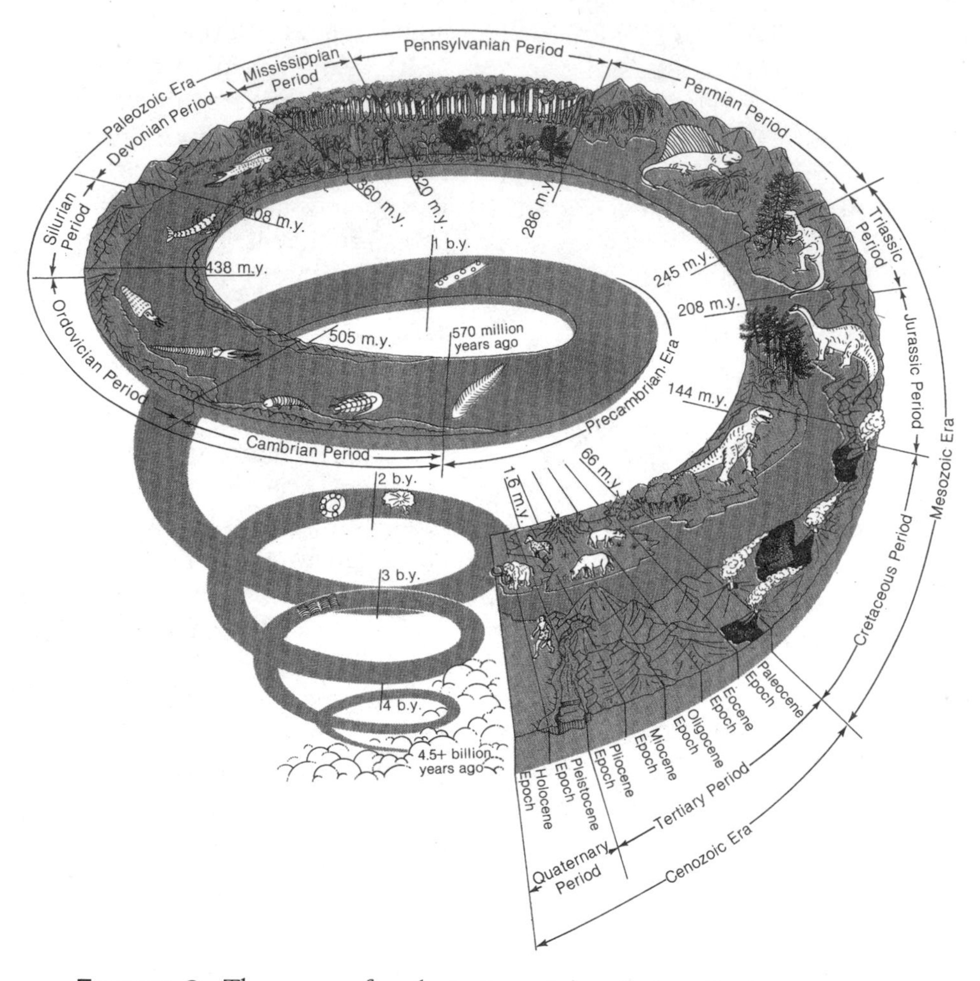

FIGURE 2 The sweep of geologic time and evolution. Geologic time is divided into eras, periods, and epochs. The earth formed 4.5 billion years ago; life began sometime in the Archean, possibly as early as 3.8 billion years ago, and exploded in the Cambrian. Dinosaurs arose in the Permian and disappeared about 160 million years (abbreviated m.y.) later at the boundary between the Cretaceous period of the Mesozoic era and the Tertiary period of the Cenozoic era. [After *Geologic Time,* U.S. Geological Survey Publication.]

The Cretaceous period began 145 million years ago and lasted until 65 million years ago, when the Tertiary began. Not surprisingly, geologists break the periods down more finely, first into epochs and then into stages. We are particularly concerned with the Maastrichtian, the last stage of the Cretaceous, which began about 75 million years ago and ended at 65 million years ago; and with the Danian, the first stage of the Tertiary, which started then and lasted until about 60 million years ago.

Walter had not gone to Gubbio to study dinosaur extinction. He and a group of American and Italian geologists were there to measure the magnetism frozen in the Cretaceous and Tertiary sedimentary rocks handsomely exposed in a deep gorge nearby. They hoped to be able to locate sections where the rocks had recorded reversals of the earth's magnetic field—times at which the north pole of the earth had acted as a south pole, and vice versa. (A magnetized rod or needle develops two poles that act oppositely. Because one end of the rod points toward the current north magnetic pole of the earth, we say that it is the north-seeking end. This property is the basis for the common compass.)

While his father, back at Berkeley, had begun to worry that physics had started to leave him behind and that his career had stalled, Walter and his co-workers were in Italy, engaged in research that no one could have expected would aid in jump-starting Luis's career. The geologists were attempting to determine the precise patterns of magnetic reversals in rocks of known age, which would then allow those same unique patterns to be used to date rocks of unknown age. Thus Walter Alvarez and his colleagues were aiming to fill in a gap in geologic knowledge, a vastly more common endeavor than launching a paradigm shift.

Geologists had discovered that, for reasons unknown, magnetic reversals were frequent (on their time scale), occurring on the average about every 500,000 years. Because all rocks of a certain age, wherever found, show the same magnetism—either normal (defined as the situation today) or reversed—we know that the reversals affected the entire earth at once. In the 1960s, analysis of the magnetic reversal patterns in rocks from the seafloor showed that sections of the floor on one side of, and parallel to, a mid-oceanic, deep-sea volcanic ridge, could be matched exactly with the pattern on the other side. Some clever scientists deduced that lavas were being extruded at these ridges and, as they cooled, took on normal or reversed magnetism, whichever was prevalent at the time. Later, the frozen lavas were dragged out to either side as the seafloor spread away from the ridge, to be replaced by a new batch of lava that, if the earth's magnetic field had meanwhile flipped, would be magnetized in the opposite direction. This proved that the seafloors diverged from ridges, and it was only a small leap to conjecture that continents, made of light, buoyant rock, would ride on top of the spreading seafloors. Thus emerged the theory of plate tectonics, the modern version of the theory of continental drift.

In order for the new magnetic time scale to be used to date rocks, it had to be tied into the standard geologic time scale that had been built up through the decades based on the diagnostic fossils contained in sedimentary rocks. To do so scientists needed to find a cross section of fossil-bearing rock of known age that had been deposited steadily and slowly, allowing the magnetic minerals in the parent sediment to capture the fine details as the earth's magnetic field repeatedly reversed itself. Gubbio was ideal. In a 400-m gorge outside the town, rocks of middle Cretaceous age, 100 million years old, are exposed at the bottom and are successively overlain by younger beds that reach well up into the Tertiary, to an age of about 50 million years. Especially prominent are thick beds of a beautiful rosy limestone—*scaglia rossa*—a favorite Italian building stone. These were exactly the kinds of rocks required by Walter Alvarez and his colleagues, for such limestones build up slowly on the deep ocean floor and their magnetism would have captured each change in the earth's magnetic field.

Not only did the team find the reversals in the rocks of the gorge, they were expressed so intricately that the geologists proposed the Gubbio section as the "type"—the world standard—for the Cretaceous–Tertiary part of the magnetic reversal time scale.[7] Walter Alvarez and his co-workers had succeeded in their effort to fill an important hole in geological knowledge. Were it not for the unique coincidence of scientific and paternal circumstances described in the Prologue, that likely would have been that.

The K–T boundary in the rocks of the Gubbio gorge can be spotted just with the naked eye (Figure 3). The white limestone below the boundary is rich in sand-sized fossils of a one-celled organism, a kind of plankton called foraminifera, many belonging to the genus *Globotruncana*. In the red limestone above the boundary, however, *Globotruncana* completely disappears, replaced by a much more scarce and much smaller foraminifer with the awkward name of *Parvularugoglobigerina eugubina*. Clearly, at this boundary something happened that killed off almost all of the "forams," as the micropaleontologists call them. Exactly at the boundary, between the two units, lies a 1-cm-thick layer of reddish clay, without fossils.

Walter brought home to Berkeley a polished specimen from Gubbio that included each of the three layers at the K–T boundary—the K, the T, and the clay in between—showed it to his father, and explained that it captured the time of the great mass extinction and marked the disappearance not only of most forams but of the dinosaurs as well. Although most nongeologists viewing this chunk of rock would have registered at most a polite curiosity (did

FIGURE 3 Luis and Walter Alvarez studying the K–T section at Gubbio, Italy. Walter Alvarez has his finger on the boundary. [Photo courtesy of University of California Lawrence Berkeley National Laboratory.]

it have potential as a paperweight?), Luis Alvarez commented that his son's description was one of the most fascinating revelations he had ever heard. Walter explained that the limestones above and below the layer contained about 5 percent clay, and suggested that perhaps the limy portion had simply not been deposited for a time, allowing the layer of pure clay to build up. Then, perhaps, lime deposition had started again, leaving behind a thin layer of clay sandwiched between two layers of limestone. Walter estimated that this might have taken 5,000 years, which would mean that the great K–T mass extinction had taken place in a mere instant of geologic time. Luis immediately proposed that he and Walter measure the length of time the clay layer had taken to form. Walter had succeeded in gaining his father's formidable attention and the game was afoot.

IRIDIUM

The magnetic reversal time scale offered one possibility for determining how much time the clay layer represented: The particular pattern of reversals above and below the clay might bracket its age of formation and allow an upper limit to be placed on how long it could have taken to deposit the layer. Alas, during this period of geologic history the reversals had not happened often enough: All that could be told is that the clay layer fell within a 6-m section of limestone deposited during a single period of magnetism, called 29 R (for *reversed*), that was known to have lasted for about 750,000 years. Six meters in 750,000 years is equivalent to 0.8 cm of sediment deposited every 1,000 years. Since the boundary clay is about 1 cm thick, at that rate it would have taken a little more than 1,000 years to form. This appeared to be an improvement over Walter's rough estimate, but since the clay is quite different from the limestone, there really was no basis for assuming the same sedimentation rate for both. The attempt to determine the time interval using the magnetic chronology thus failed, but in another way the effort succeeded, for the mind of Luis Alvarez was now locked in.

What was needed, he reasoned, was a geologic clock that had been operating at the time the clay layer formed but that could be read today. Because no one knew how much time the clay layer might represent, the clock might have to measure small differences. None of the standard geologic clocks—the ones based on radioactive parent-daughter pairs of atoms that are used to calculate exact ages—had enough sensitivity or would work on the chemical elements in the clay layer. Therefore, as he had done so many times in his career, Luis Alvarez invented a new technique. To do so, he looked not down to the earth but up to the heavens, postulating that the amount of a rare metallic element called iridium might provide the clock.

When the earth formed, iridium, like other elements of the platinum group (which includes osmium, palladium, rhodium, and ruthenium), accompanied iron into the molten core, leaving these elements so rare in the earth's crust that we call some of them precious. Their abundance in meteorites and in average material of the solar system is many times higher than in the earth's crust. The iridium found in sedimentary rocks (and often it is too scarce to be detected) appears to have settled from space in a steady rain of microscopic fragments—a kind of cosmic dust—worn from tiny meteorites that form the shooting stars that flame out high above the earth. Such meteorites are believed to reach the upper atmosphere at a constant rate, so that the metallic rain falls steadily to earth, where it joins with terrestrial material—dust eroded from the continents and the skeletons of microscopic marine animals—

to settle to the bottom of the sea. There it is absorbed into the muds that accumulate on the seafloor and that eventually harden into rock.

But over geologic time, the rate of accumulation varies greatly. Since one component, the meteoritic, is arriving at a constant rate from space and the other, the terrestrial, is accumulating at a varying rate, the percentage of meteoritic material in a deep-sea sedimentary rock provides a gauge of how fast the terrestrial component built up: The greater the percentage of meteoritic debris in a given thickness of rock, the slower the sediment accumulated, and vice versa. The rate at which meteorites fall on the earth is known, as is the amount of iridium in meteorites, so that the iridium content of sediments can be used as proxy for the total amount of meteoritic material they contain. Luis Alvarez later discovered that two scientists from the University of Chicago had tried to use iridium in this way to measure sedimentation rates but without success. "Fortunately, I hadn't heard of their work," Luis commented of the Chicago scientists, "If I had, I'm sure we wouldn't have bothered to look for iridium at the K–T boundary."[8]

The samples necessary for testing the iridium clock were readily available from the Gubbio clay layer, but measuring the expected low levels of iridium required a research nuclear reactor. Fortunately, the Alvarezes and their Berkeley colleagues Frank Asaro and Helen Michel had access to one. The reactor allows neutrons to bombard a rock sample and cause atoms of an isotope of iridium to become radioactive and to emit gamma rays of a distinctive energy. The number of such rays emitted per second is counted and is proportional to the amount of iridium in the original material. When analyzing at the parts per trillion level, however, it is extremely difficult to eliminate contamination (from the iridium that is always present in platinum jewelry, for example).

Walter Alvarez selected samples ranging over the Gubbio section—above, below, and at the K–T boundary—and brought them back to Berkeley for analysis. Samples from above and from below the boundary had the predicted amounts of iridium, about the same as had been measured by others in deep-sea clays—300 parts per trillion (ppt) or so. The samples from the boundary clay, however, revealed an earthshaking surprise—iridium levels 30 times higher than those in the limestones on either side! Back-of-the-envelope calculations showed that Walter's original idea—that the clay had built up when the limestone for some reason ceased to be deposited—could not be correct because then the clay would have taken an impossibly long time to form. Thus the attempt to use iridium as a clock failed, but, as often happens in science, no sooner does one idea fall by the wayside than another springs up.

To check that the startling result was not somehow a bizarre characteristic of the K–T boundary clay at Gubbio, the team analyzed two clay layers contained within the limestones above and below the boundary and found both to have low levels of iridium. Thus the iridium anomaly was associated with the thin K–T boundary clay, not generally with clays from the Gubbio region.

If not an anomaly of Gubbio clays, perhaps the iridium spike was merely a local aberration. To find out, the scientists needed to find another site where the K–T boundary is exposed, collect samples, and analyze them for iridium. As an indication of just how little was known about the K–T boundary in the late 1970s, even a knowledgeable geologist like Walter had no idea where to look. As would any intelligent person in a similar quandary, he went to the library. There he discovered a reference to the sea cliffs south of Copenhagen, which contain a classical and thoroughly studied K–T rock section where, as in Italy, a clay layer marks the precise boundary. Measurements of the amount of iridium in the Danish clay by Frank Asaro showed an even greater enrichment—160 times background. The Alvarez team was clearly onto something: The iridium anomaly was not restricted to Italian rocks and might even be a worldwide phenomenon.

What did the Alvarezes know at this point? That at two widely separated sites, abnormally high levels of one of the rarest metals in the earth's crust occur in the exact thin layer that marks the great K–T extinction and the demise of the dinosaurs. They concluded that this could hardly be due to coincidence—the high iridium level must somehow be linked to the extinctions. They knew that finding how this linkage had occurred was the 65-million-year question; if they could, they might solve the age-old riddle of dinosaur extinction.

Since iridium is many times more common in meteorites and in the solar system in general than in crustal rocks, the Alvarezes began to consider extraterrestrial sources for the Gubbio iridium. The first idea they pursued was the one paleontologist Dale Russell favored. Exploding stars, or supernovae, which generate and then blast cosmic material throughout the galaxy, might have implanted the K–T iridium, suffusing the earth with deadly cosmic rays and thus causing the extinction. Such nuclear furnaces give birth to a wide variety of chemical elements, including plutonium. One isotope of plutonium, Pu 244, is a diagnostic marker of supernovae explosions. A diligent search for Pu 244 in the boundary clay came up empty, however, so the scientists had to abandon the supernova theory.

A Berkeley colleague, Chris McKee, suggested that a large asteroid could have provided the iridium. This made sense, for iridium is present at high levels in lunar soils, where it has presumably been emplaced by impacting meteorites. For many months, however, the Alvarez team was unable to figure out how the impact of an asteroid at one spot on the earth's surface could have caused a mass extinction everywhere. How did the effects get spread around the globe? Luis later recalled that he had invented a new scheme a week and shot each down in turn.

The critical clue arose in a way that further illustrates the strong scientific ties within the Alvarez family. In 1883, the island volcano Krakatoa, in the Sunda Strait between Java and Sumatra, blasted itself to pieces in one of the most violent eruptions of modern times, scattering debris as far as Madagascar. People 5,000 km away heard the explosion. Walter Alvarez, Sr., the physician father of Luis, had given his son a volume describing the Krakatoa event published by the Royal Society of London in 1888; Luis in turn had passed it on to the younger Walter. Now Luis asked for the volume back so that he could study the consequences of a dust-laden atmosphere. The Royal Society volume estimated that the Krakatoa explosion had blasted 18 km^3 of volcanic material into the atmosphere, of which about 4 km^3 reached the stratosphere, where it stayed for more than two years, producing some of the most remarkable sunsets ever witnessed. (In comparison, the eruption of Mount St. Helens in 1980 is estimated to have released about 2.7 km^3 of volcanic rock; the eruption that formed the giant Yellowstone crater about 3,000 km^3.)

Krakatoa caught Luis's attention, and he proposed by analogy that 65 million years ago a large meteorite struck the earth and sent up such a dense cloud of mixed meteoritic and terrestrial debris that it blocked the sun. This successively caused world temperature to drop, halted photosynthesis, choked the food chain, and led to the great K–T mass extinction and the death of the dinosaurs. Luis and his Berkeley gang phoned Walter, then in Italy, to announce their exciting conclusion and to propose that the idea be presented at an upcoming meeting on the K–T boundary in Copenhagen, which both Alvarezes could attend. Although Luis was anxious to explain to paleontologists the cause of dinosaur extinction, Walter knew better and urged him to stay home.[9] While the physicist and his chemist colleagues did remain in Berkeley, Walter journeyed to the Copenhagen meeting. There he met Dutch geologist Jan Smit, the only other person present to give any credence to the embryonic theory.

Luis, Walter, Frank Asaro, and Helen Michel spent the next months preparing a long paper describing their theory, which they then submitted to *Science*. Its editor, Philip Abelson, had been Luis's graduate student at Berkeley in 1939 and a longtime colleague. Perhaps having grown weary of dinosaur extinction theories, Abelson responded that the paper was too long and that furthermore, since *Science* had published many papers purporting to solve the mystery of the dinosaurs, "at least n–1 of them must be wrong"—a scientist's way of saying that only one could be right. The authors submitted a shorter version (still twice as long as the journal's typical lead article); it appeared in the issue of June 6, 1980.[10]

COSMIC WINTER

When presenting a theory far outside the mainstream, the first question is whether it is credible. The burden of establishing credibility properly rests with the proposers; if they are unable to do so convincingly, the theory is best let lie. A theory whose credibility has been weighed and found wanting may not have been proven false, but the finding does serve to direct research elsewhere. On the other hand, as we shall see, too often in geology a magisterial authority has made a pronouncement—that the earth can be no more than 20 million years old, that continents cannot move, that few or no terrestrial craters could have been formed by meteorite impact—and later been found to be dead wrong, costing decades of fruitful research. It is important not to pursue every offbeat idea but equally important not to draw conclusions too hastily. Yesterday's offbeat notion has often become today's paradigm. Arthur C. Clarke caught the proper spirit when he said that "if an elderly but distinguished scientist says that something is possible he is almost certainly right, but if he says that it is impossible he is very probably wrong."[11]

One way for the Alvarezes to test the credibility of their theory was to estimate the size of the alleged impactor. If it turned out to be as large as, say, Mars, or as small as the tiny meteorites that give rise to shooting stars, credibility would be undermined. A meteorite the size of a planet cannot have hit the earth 65 million years ago or all life at the surface would have been eradicated—nothing would have made it through. At the other extreme, tiny shooting stars burn up in the atmosphere and thus have no effect. To be credible, the size of the putative impactor would have to be much smaller than a planet and much larger than a shooting star.

The size of the alleged K–T meteorite could no longer be measured directly—the impact explosion would have blasted it to pieces. But what of the residue it might have left behind, the iridium in the Gubbio clay layer? If the impact event had worldwide effects, approximately the same amount of iridium as found at Gubbio would have been deposited in a layer that extended all around the earth, coating its entire surface. Knowing both the amount of iridium in the Gubbio clays and the size of the surface area of the earth, the Alvarez team calculated that about 200,000 tons of iridium had been emplaced. Since they knew the average iridium content of meteorites, they were then able to figure out how large a meteorite would have been required to deliver that much iridium. Using reasonable assumptions as to density and shape, the answer was a meteorite about 6.6 km in diameter. Applying the same technique to the Danish clays gave about 14 km. That the two estimates agreed within about a factor of 2 was encouraging at this rough level of calculation. Averaging them gave 10 ± 4 km, neither as large as a planet nor as small as the pip-squeaks that produce shooting stars, and well within the credible range. The figure of 10 km has become accepted as the diameter of the Alvarez impactor. That happens to be about the elevation of Mt. Everest, the earth's highest mountain. Imagine that Everest, instead of standing above the already lofty Himalayan plateau, rose straight from the sea to its height of over 29,000 feet. Now imagine a mountain of that size approaching the earth at a speed of 100,000 miles per hour. No thanks!

The Alvarezes next compared a meteorite 10 km in diameter with three observational facts:

1. Asteroids (solid rocklike meteorites) and comets (balls of dirty ice), either of which could have produced the impact, in the range of 5 km to 10 km in diameter are relatively plentiful in space and are routinely observed through telescopes.

2. Estimates based only on astronomical observations show that an asteroid or a comet 10 km in diameter should strike the earth about every 100 million years, so having one hit 65 million years ago but none since would fit the observations (see Table 1).

3. Over 150 terrestrial impact craters are known; from their size and frequency, crater experts estimate that a 10-km object

strikes about every 100 million years. This conclusion, based only on known craters, is completely independent of the one based on astronomy, yet it gives the identical result.

Thus the Alvarez impact theory described an event that is rare but that does occasionally take place and when it does, must produce large-scale effects. Even though in the early 1980s geologists were still coming to understand the role of impact in the history of the solar system, the Alvarez theory was within the range of what was known and observable. It clearly passed the credibility test and needed to be taken seriously.

The theory itself consisted of two parts: first, that a meteorite struck the earth 65 million years ago, and second, that the effects thus produced were so severe that they led to the K–T mass extinction. Unfortunately for their theory, but fortunately for *Homo sapiens,* it is not easy to test the second part, for no large meteorite has struck in the minute fraction of geologic time recorded by human history.

One approach to the problem of verifying the theory's second claim is through computer modeling. In 1983, influenced by the Alvarez theory, a group of scientists that included the late Carl Sagan used computer models to show how a nuclear war in which fewer than half of the combined number of warheads then available to the United States and the Soviet Union were exploded would throw enough dust, smoke, and soot into the atmosphere to block sunlight for several months, particularly in the Northern Hemisphere. This might set in motion the same sequence of events as predicted by the Alvarezes (lowering temperatures by tens of degrees, halting photosynthesis, destroying plant life, and disrupting the food chain). The ozone layer might also be affected, allowing the sun's ultraviolet radiation to penetrate and cause further damage. Their paper, which appeared in *Science,* concluded that nuclear war would have so few survivors, if any, that it would produce another great extinction—this time, possibly of *Homo sapiens.*[12] The threat of nuclear winter caught the attention of the world and may have been influential in halting the growth of nuclear weapons and ending the Cold War.

Though it would not include deadly radioactive fallout, cosmic winter would be far worse than nuclear winter. The impact of a 10-km meteorite would release a vastly greater amount of energy than Krakatoa, which caused the death of 35,000 people. It would do far more damage than the atomic bomb that was dropped on Hiroshima, which had the energy equivalent of about 13 *kilo*tons of TNT. Traveling at 25 km/sec or more, the mountain-sized meteorite would strike the earth with the force of 100 *million megatons* of TNT (10^{14} tons; 10 followed by 14 zeros), more than 7 billion times as much energy as the bomb dropped on Hiroshima—in fact, vastly more energy than the explosion of *all* of the 60,000 nuclear weapons that existed at the height of the Cold War. To comprehend the power of meteorite impact, try to imagine the simultaneous explosion of 7 billion bombs like the one dropped on Hiroshima—one for every person on earth and 10 for every square kilometer of the earth's surface. The terrorist bomb that destroyed the Alfred P. Murrah Federal Building in Oklahoma City in 1995 had an energy equivalent measured not in megatons, not in kilotons, but in tons—2.5 *tons*. The K–T impact was 40 trillion times larger. The dinosaurs, concluded the Alvarezes, never had a chance.

LOSING BY A NOSE

Even today, in the era of electronic mail, faxes, and international flight, it is still possible for two individuals or groups to work independently, unknown to each other, and to come to the same conclusion simultaneously. This nearly happened in the case of the meteorite impact theory.[13] In 1974, geologist Jan Smit began to study the K–T boundary at Caravaca, Spain, for his doctoral dissertation, focusing on the disappearance of the microscopic foraminifera there. At the start, he thought their sudden exit at the boundary was only apparent, caused by an erosional gap that made a gradual extinction appear falsely sharp. But the sudden extinction persisted even in sections without visible gaps. Smit decided to see whether there were invisible, chemical changes.

In the spring of 1977, he sent off to the Dutch interuniversity laboratory in Delft a set of 100 K–T samples for neutron activation analysis, asking the scientists there to determine the concentrations of various elements. Because there was no reason for him to do so, Smit did not include iridium on his list of elements to be studied. When the results came back, the thin boundary clay turned out to have concentrations of nickel, cobalt, chromium, arsenic, antimony, and selenium that were orders of magnitude higher than in the limestones on either side. Smit published his finding that the extinctions were rapid, and, as he describes it, "began to speculate about extraterrestrial causes."[14] What Smit did not know was that the iridium levels of his samples had actually been available in the analysis records, but, since they had not been requested, were not reported to him. On that may have hung the priority for the meteorite impact discovery—otherwise we might be discussing the Smit theory rather than the Alvarez theory. Serendipity thus can also work in reverse: A scientist may be unlucky and miss having the chance to make a critical observation.

Two years later, after a lengthy bout with mononucleosis, Smit was bowled over to read in the *New Scientist* that scientists at Berkeley had discovered high iridium concentrations in the Gubbio boundary clay.[15] Smit sent his Caravaca samples to Jan Hertogen in Belgium, who had the equipment to analyze them for iridium. At 28,000 ppt, iridium in the Caravaca K–T boundary clay turned out to be five times higher than at Gubbio! This prompted the head of the Neutron Activation Department at the interuniversity laboratory in Delft to go back over the archived data from the earlier analysis—the unreported iridium peak came in at 26,000 ppt.

Smit and Walter Alvarez met at the Copenhagen K–T conference in September 1979 and, finding themselves alone in giving credence to the meteorite impact theory, soon became fast friends, and they have remained so. At first, Smit preferred the supernova explanation for the K–T mass extinction, but subsequent discussions with an astronomer colleague soon convinced him that the iridium levels were too high. In December 1979, he received a preprint of the paper that the Alvarez team had submitted to *Science.* One month later, he and Hertogen submitted a paper to *Nature* based on their Caravaca findings, noting that "the impact of a large meteorite may have provided the iridium" and caused the K–T mass extinction.[16] The paper appeared in May 1980, one month before the original Alvarez paper. Thus Smit got into print first, which would have allowed a less scrupulous person to claim priority. He knew, however, that the chronology of events required that he give credit for the discovery to the Alvarezes, which he did. But it was a near thing. And Smit had not only supernovae, but meteorites, on his mind. But in science, as an excellent practitioner like Smit knows full well, there is no second prize.

Three Princes of Serendib

The Alvarez team did not proceed according to the stereotype of the scientific method: They did not hypothesize that the dinosaurs were killed by the effects of a meteorite impact, reason out that iridium would provide the evidence, and then set out to test their theory by measuring iridium levels in the K–T boundary clay. Rather, while investigating a completely different idea—that iridium could be used to measure sedimentation rate—they discovered the iridium "spike." This is often how science works: While looking for one thing, sometimes for nothing, a scientist by accident makes an important discovery. In the eighteenth century, Sir Horace Walpole read a fairy tale about the "Three Princes of Serendib" (Sri Lanka), who "were always making discoveries, by accidents and sagacity, of things which they were not in quest of," and he coined the term *serendipity* to describe their approach. Royston Roberts' delightful book of that name describes some of the many discoveries, aside from that of dinosaur extinction, that have had their origin in accidents: penicillin, X rays, Teflon, dynamite, and synthetic rubber, to name a few.[17]

Accidents happen to everyone, the great and the not-so-great alike, but accident does not necessarily imply serendipity. The Alvarezes made an accidental discovery, but turned it serendipitous by what they did next. They could have put down the unanticipated finding of high iridium levels to contamination or to a freak event and ignored it. Instead, they immediately turned their attention to finding out why the strange result occurred; that led them on to earthshaking discoveries.

In the absence of Pasteur's "prepared mind," chance turns away, accidents are not converted into serendipitous discoveries, and average scientists are sorted from great. The minds of most geologists, trained to believe that the earth changed slowly and imperceptibly over geologic time, certainly were not prepared to accept the meteorite impact theory. Not only was the introduction of the theory unnecessary, it appeared to many geologists to be a misguided attempt by outsiders to reverse 150 years of progress.

Chapter 2

The Past as Key to the Present

A science which hesitates to forget its founders is lost.[1]
Alfred North Whitehead

Resistance

After Luis Alvarez informed a physicist colleague that the absence of Pu 244 in the boundary clay negated the supernova theory, he received the reply: "Dear Luie: You are right and we were wrong. Congratulations." To Luis, this response "exemplified science at its best, a physicist reacting instantly to evidence that destroys a theory in which he previously believed."[2] He could never understand why the paleontologists did not react the same way.

He could have taken a lesson from the theory of continental drift, which took decades to find acceptance among geologists. Early in the nineteenth century, mapmakers and others had noted that the coastlines of South America and Africa fit together like two pieces of a jigsaw puzzle. In 1918, German meteorologist Alfred Wegener extrapolated from this apparent coincidence to develop a full-fledged theory, backed by a volume of geological evidence, that held that continents are not fixed in place on the surface of the globe, but drift about, colliding, welding together, and sometimes separating along a new fracture. Each single piece of evidence that Wegener presented, however, was circumstantial and therefore could be ascribed to coincidence. Furthermore, he could present no plausible mechanism to explain why the continents should have moved. His idea failed to catch on and came to be regarded, at least by American geologists, as having been falsified to the point of being laughable. The most widely used American textbook on earth history during the 1950s did not contain the words "continental drift." A few years later, but before the plate tectonic revolution, the great Canadian geologist J. Tuzo Wilson lectured on continental drift at MIT. The attitude of the faculty (and therefore of most students) was to regard Wilson, a man of impressive dignity, rather like an eccentric uncle—still a member of the family, but not to be taken seriously. Within a few years, however, Wilson was not only vindicated, but rightfully hailed as a hero of the revolution.

Ursula Marvin, professor of astronomy at Harvard, recounted a similar episode when in 1964 she proposed that geologist Robert Dietz (like Wilson, a courageous visionary) be invited to Harvard to speak about his remarkable theory that meteorite impact created the enormous body of igneous (once-molten) rock in Sudbury, Ontario.[3] It is one of the world's greatest, and most studied, sources of nickel ore. Among geologists, it was famous as well as mystifying, for, as with dinosaur extinction, none of the many theories that had been proposed for its origin had received general acceptance. Nevertheless, the Harvard graduate students who had worked at Sudbury "staged a boycott . . . intuitively reject[ing] impact as unworthy of their magnificent structure and, indeed, a *deus ex machina* appropriate only to science fantasy."[4]

In resisting the Alvarez theory, geologists were merely behaving as people have done throughout history. One who has invested time and effort—possibly an entire career—based on the notion that the continents are fixed, or that geologic change is slow and unaffected by cosmic events, when confronted with an entirely different idea, has to cast off years of work and assumptions and possibly even renounce previously published conclusions. Max Planck, the father of quantum mechanics and a Nobelist, summed it up: "An important scientific innovation rarely makes its way by gradually winning over and converting its opponents. . . . What does happen is that its opponents gradually die out and that the growing generation is familiar with the idea from the beginning."[5]

Another reason that people resist new theories is because the authorities in a field, who by definition have achieved their eminence working within the prevailing paradigm, often say the new theory is impossible, or at least, highly unlikely. After all, if the new theory is correct, then one of theirs may need replacing. In the nineteenth century the great physicist Lord Kelvin, unaware that radioactivity existed to provide a source of heat, pronounced that the earth must have been hot initially and been cooling ever since. Working backward, he calculated that the earth could be no more than 20 million years old. His eminence caused this erroneously short time scale to be accepted, delaying for decades recognition that the true extent of geologic time is on the order of 4.5 *billion* years. (Kelvin also denounced X rays as a hoax.) G. K. Gilbert, chief geologist of the U.S. Geological Survey and the leading geologist of his day, incorrectly concluded that Meteor Crater, Arizona, was not formed by meteorite impact, leading to a dogma that was decades in the unmaking. When the deans of American geology and the faculties of research universities scoffed at the theory of continental drift, budding geologists of the 1950s and early 1960s chose other topics. The history of science is full of the undue influence of magisters—authoritative masters—whose pronouncements receive an uncritical acceptance.

WITHOUT HELP FROM A COMET

Walter Alvarez could have told his father that it is hard to find any idea in the history of science more consistently and continuously spurned by authorities than the notion that meteorite impact has in any way affected the earth. The rejection stretches back to the dim beginnings not only of geology but of science itself. In the 1680s, William Whiston, mightily impressed by the great comet that had opened that decade, wrote that God had directed the comet at the earth and that its impact had produced both the tilt of the earth's axis and its rotation, had cracked the surface, and had allowed the waters to rise to create the biblical flood.[6] Even though Whiston convinced no one, his idea so offended Charles Lyell, a founding father of geology, that nearly 150 years later he went out of his way to debunk Whiston's suggestion, writing that he had "retarded the progress of truth, diverting men from the investigation of the laws of sublunary nature, and inducing them to waste time in speculations on the power of comets to drag the waters of the ocean over the land—on the condensation of the vapors of their tails into water, and other matters equally edifying."[7] Lyell's disciple, Charles Darwin, was equally convinced that catastrophes played no part in earthly events, writing: "As we do not see the cause [of extinction], we invoke cataclysms to desolate the world, or invent laws on the duration of the forms of life."[8]

A century later, little had changed. E. H. Colbert of the American Museum of Natural History, the dean of American dinosaur studies at the time, wrote: "Catastrophes are the mainstays of people who have very little knowledge of the natural world; for them the invocation of a catastrophe is an easy way to explain great events."[9]

Walter Bucher was an eminent American geologist and the country's leading authority on mysterious rock structures that geologists referred to as "cryptovolcanic." Here and there around the globe, rocks at the surface are broken into a set of concentric faults that form a bull's-eye pattern. Often the structures occur in sedimentary rocks hundreds of miles from the nearest volcanic lavas. Nevertheless, geologists, looking down and not up, could think of no other plausible origin for these structures than that they were created by gases exploding from invisible underground volcanoes. A few brave souls had the temerity to suggest that these features might actually *be* a kind of bull's-eye, marking the target struck by an impacting meteorite. In 1963 Bucher wrote in the definitive paper rebutting this view: "Before we look to the sky to solve our problems miraculously in one blow, we should consider the possibility that cryptoexplosion [cryptovolcanic] structures and explosion craters may hold important clues to processes going on at great depth below our feet, even if it threatens to lead us back to another 'traditional' concept, that of cooling of the outer mantle. Distrust in traditional thinking should not deter us from looking hard at all aspects of the problem. Doing so will probably yield more useful results than computing possible velocities of imagined meteorites."[10] In other words, do not try to solve geologic problems by appealing to missing meteorites from space.

Tony Hallam, a distinguished British geologist, advised that "Environmental changes on this planet as recorded by the facies [rock types] should be thoroughly explored before invoking the *deus ex machina* of strange happenings in outer space. . . . It is intuitively more satisfying to seek causes from amongst those phenomena which are comparatively familiar to our experience."[11] A 1986 review article intended to sum up matters for students and teachers stated that "it is not necessary to invoke a meteorite impact to explain the K/T extinctions, and, in actuality, an impact does not explain those extinctions."[12]

The award for the most unlikely source of negative reaction goes to the *New York Times,* which in a 1985 editorial curiously titled "Miscasting the Dinosaur's Horoscope," declared that "terrestrial events, like volcanic activity, or change in climate or sea level, are the most immediate possible cause of mass extinctions. Astronomers should leave to astrologers the task of seeking the causes of earthly events in the stars."[13] This prompted Stephen Jay Gould to fantasize in *Discover* magazine what might have been written in *Osservatore*

Romano of June 22, 1663: "Now that Signor Galileo, albeit under slight inducement, has renounced his heretical belief in the earth's motion, perhaps students of physics will return to the practical problems of armaments and navigation, and leave the solution of cosmological problems to those learned in the infallible sacred texts."[14]

The attempt to debunk the Alvarez theory was not the first time the *New York Times* recommended that scientists come to their senses and follow its advice. In a 1903 editorial, the paper advised aviation pioneer Samuel Pierpont Langley "not to put his substantial greatness as a scientist in further peril by continuing to waste his time and the money involved in further airship experiments. Life is too short, and he is capable of services to humanity incomparably greater than can be expected to result from trying to fly."[15] How fortunate that neither Langley nor the Alvarezes paid any attention.

When an author feels compelled to exorcise predecessors dead for over 100 years, when the paper that publishes "All the News That's Fit to Print" feels entitled to weigh in on its editorial pages, when theories are judged not on whether they meet scientific tests but on whether they are required or satisfying, it is obvious that the suggestion that earthly events have extraterrestrial causes leads otherwise sober-minded folks to give sway to their emotions. But why does an appeal to factors outside the earth produce such a negative reaction? Geologists, at least, have a reasonable answer: Starting with their first course in the subject, they have been taught that the earth simply does *not* change in response to sudden catastrophes. This notion of geology by catastrophe was disproven a century and a half ago; to resurrect it in the late twentieth century would be to return the science to its prescientific days.

No Powers Not Natural to the Globe

The key concept underpinning the geologists' view that slow change can accomplish everything is the vastness of geologic time. At some point in the not-so-distant future, 1,000 years or 10,000 years from now, when fossil fuels and valuable ore deposits are gone, the permanent contribution of geology surely will be the concept of the limitless extent of geologic time—what John McPhee aptly calls "deep time." This is a major intellectual contribution equivalent to that of astronomy: the realization that the earth is not the center of anything, rather it is an inconspicuous planet revolving around one star among billions of stars, in one galaxy among billions of galaxies. But although we can observe other planets, stars, and galaxies, a human lifetime is so short that deep time surpasses understanding. We can fathom a few hundred years, even a few thousand, but we cannot comprehend the passage of millions and billions of years. A metaphor that well captures the different character of geologic time, again from McPhee, uses the old English yard, the distance from the King's nose to the tip of his extended finger, as the equivalent of geologic time. Apply to the regal digit one light stroke of a nail file, and the equivalent of human history disappears. Comprehension of geologic time must be accorded its rightful place as one of the great achievements of human induction. Its importance to an analysis of meteorite impact theory is that with time enough—with "time out of mind"— earth history can be fully explained with no need to appeal to catastrophes. To do so is to betray the key success of geology: recognition that within the vast length of geologic time, everything could be accomplished.

Because it is so foreign to our human time scale, the concept of deep time not surprisingly took several centuries to develop. Leonardo da Vinci was among the first to realize that the fossil shells found high in the mountains were the remains of animals that had once lived deep in the sea. Had he not been such a great painter, we would likely remember Leonardo as the outstanding scientist of his day. Another great advance came in the middle of the seventeenth century, when a Dane named Nicolaus Steno compared fossils encased in rock with the shark's teeth that sailors had brought him for study. He could see that the two were identical and reasoned that the teeth had somehow become enclosed in the rock; the teeth had existed before the rock had fully formed, therefore the teeth were older. This point seems elementary now but in its day was revolutionary, for it led Steno to realize that some materials of the earth were older than others and therefore that the earth, like a person, has a history that can be interpreted and understood.

Until two centuries ago, science was required to be consistent with the Book of Genesis. In the seventeenth century, James Ussher, archbishop of Armagle, working backward from the beginning of that book, allowing due time for the events described, calculated that the earth had been formed about 6,000 years before. Since all of the earth's history had to be fitted into such a short period, in this view, geologic processes must be rapid and catastrophe must be the rule. (Creationists today argue that the earth can be no more than 10,000 years old, even though the Sumerians were so advanced as to have a written language 6,000 years ago.)

But by the 1780s some had come to find catastrophism untenable, for it did not agree with the slow, inexorable processes of erosion and deposition that they observed and analyzed. One such person was Scotsman James Hutton. A devout man, he believed that God had created the earth for the express benefit of mankind, and, since he could see the earth wearing away, became convinced that some process must restore it. Otherwise the continents would steadily erode into broad, uninhabitable plains—surely not what God had intended. Hutton sought and found evidence that sediments worn from the continents and deposited in the sea are subsequently hardened, heated, uplifted, and returned to the continents to start the process all over again. He viewed earth history as a series of endless cycles of decay and rejuvenation, with, in his most famous phrase, "no vestige of a beginning—no prospect of an end." His cycles required vastly longer periods of time than allowed by a strict interpretation of the Bible; indeed, they implied an "abyss of time."

It is curious that Hutton has wound up as the "founder of geology," for he started with theology rather than with the rocks, drew conclusions first and then sought evidence for those conclusions, and propounded a theory of endless cycles that is at best a vast oversimplification. Today we would hardly regard these as the mark of a great scientist. But Hutton has established his place in the pantheon of geology not for these reasons but because he enunciated a principle that was to become central to geologic thought and practice: "Not only are no powers to be employed that are not natural to the globe, no action to be admitted of except those of which we know the principle, and no extraordinary events to be alleged in order to explain a common appearance . . . we are not to make nature act in violation to that order which we actually observe . . . chaos and confusion are not to be introduced into the order of nature, because certain things appear to our partial views as being in some disorder. Nor are we to proceed in feigning causes, when those seem insufficient which occur in our experience."[16]

In other words, in explaining the earth, we are to call upon only those processes that we observe. Given time enough, they will do the job. This principle was to become the core concept of geology. Hutton summed up its central premise, in a phrase learned, if not thoroughly comprehended, by every beginning student of geology since: "The present is the key to the past." Like most slogans, this one has a deceptively simple appeal. A moment's thought reveals that since the time scale of human history is so short compared to deep time, important processes that act only rarely could have occurred long ago, but never since, so that there has been no chance for us to observe them. To the extent that they have not been seen, the present is not the only, and certainly not the complete, key to the past. How was this seemingly obvious point ignored? Largely because of the influence of Charles Lyell.

Born in 1797, the year that Hutton died, Lyell, through his *Principles of Geology*, became the most influential geological writer of all time.[17] He was a lawyer who knew how to frame an argument, and his treatise, presented as a textbook, was in fact a "passionate brief for a single, well-formed argument, hammered home relentlessly," as Stephen Jay Gould has described it.[18] Lyell believed with Hutton that God designed the earth for human beings, but that once He set the earth on its path, He never again intervened in its workings. Natural laws were invariant. The processes that we observe today, and only those processes, have been in operation since the beginning. Natural law and process are constant. In Lyell's philosophy there were no more things in heaven than there were on the earth; he needed no "help from a comet" to explain earthly processes.[19]

Lyell also believed that the rate at which geologic processes acted was constant. He wrote, "If in any part of the globe the energy of a cause appears to have decreased, it is always probable that the diminution of intensity in its action is merely local, and that its force is unimpaired, when the whole globe is considered."[20] He meant that violent upheavals in one part of the earth are offset and averaged out by quiescence elsewhere, leaving the overall rate of change over time the same. Drastic change therefore, like all politics, is local.

Given Lyell's belief that neither natural law, the kinds of processes that affect the earth, nor their rate, ever change, it is not surprising to find that he also believed that the earth has always looked as it does now, its history revealing no evidence of directional change. The pterodactyl is gone, true, but when climatic conditions are again favorable, it may return to "flit again through the umbrageous groves of treeferns."[21] The earth always remains in the same state, neither progressing nor deteriorating.

We can divide Lyell's thesis into the constancy of law and process, and the constancy of rate and state. In what Gould has called "the greatest trick of rhetoric . . . in the entire history of science,"[22] Lyell gave them all the same name—*uniformity*—thus obscuring the fundamental difference between them for well over a century. William Whewell, who reviewed the second volume of Lyell's book, lumped his two meanings together under the unwieldy term *uniformitarianism*. (He also coined the word *scientist*.) Whewell asked whether "the changes which lead us from one geological state to another have been, on a long average, uniform in their intensity, or have they consisted of epochs of paroxysmal and catastrophic action, interposed between periods of comparative tranquillity?"[23] He predicted that the question "will probably for some time divide the geological world into two sects, which may perhaps be designated as the Uniformitarians and the Catastrophists."[24] He was both wrong and right. The biblical catastrophists of Lyell's day were clearly in the wrong and disappeared more quickly than Whewell predicted, but they have been replaced by today's neocatastrophists, the pro-impactors.

In order to understand how the earth works, and how geologists practice their science, the two types of uniformity have to be disentangled. The constancy of law and process, which Gould has called *methodological uniformitarianism*, describes not how the earth works, but how geologists ought to work. In common with other scientists, geologists reject supernatural explanations and employ known, simple processes before they turn to unknown, complicated ones. For example, today we can see streams eroding and depositing; it is only logical to assume that they have been doing so ever since liquid water appeared on the surface of the earth, and that many sedimentary deposits were formed by stream action.

Of course, this is not only the way that all science ought to work and does. It is nothing more than common sense, well expressed by William of Ockham in the fourteenth century: "One should not assume the existence of more things than are logically necessary." Throughout the history of science up to the space age, meteorite impact was simply a vague idea with very little to support it. Thus to endorse it, based on the knowledge available, was to violate Ockham's razor, as it has come to be called. But today just the opposite is the case. As we will see, scientists today are logically required to acknowledge that impact has happened numerous times.

All scientists reason from cases in which they can examine cause and effect to those in which only effect is evident. This is especially true in geology, where practically everything took place before we arrived. But nothing is special about methodological uniformity; it says only that geology is a science.

The classic example of the success of methodological uniformitarianism was the work of Swiss geologist and naturalist Louis Agassiz, who noted that modern glaciers high in the Alps could be seen to gouge rocks from their beds and to carry the dislodged pieces along, sometimes moving boulders as big as a carriage. When these glaciers melted back, the rocks over which they had passed were left polished and grooved; as they receded from their points of farthest advance, they were seen to leave behind ridges of rock debris. Agassiz then proceeded to find all of these features and more down the Alpine valleys, far below the snouts of present glaciers. He reasoned that glaciers must once have extended over a much greater range than they do today and proposed that there had once been a great ice age. This was not so hard to imagine when looking up a Swiss valley, but other scientists, finding glacial deposits far below the furthest extent of present ice sheets, extended the reasoning to conclude that huge ice sheets as much as a mile thick had advanced over much of the Northern Hemisphere.

Lyell presented the other type of uniformity—of rate and state, which Gould has called *substantive uniformitarianism*—as an a priori description of the way the earth works: Over the long span of earth history there has been no directional change, no progression. But substantive uniformity was tested and falsified in Lyell's own century, when it was learned that glaciers of vast size had advanced over the continents, that the seas at times had risen to drown the land and at other times had dried up, that mountain ranges had risen and been eroded away. Clearly, processes have operated at different rates and the earth has changed. The coup de grace to substantive uniformitarianism was the obvious progression shown by the fossil record, leading from one-celled bacteria in Precambrian rocks to modern *Homo sapiens.* But Lyell accepted evolution only in the 1866 edition of his *Principles,* and only then, Gould believes, because "it permitted him to preserve all other meanings of uniformity."[25] Since he believed the rate of biological change always to be the same, Lyell was forced to conclude that the vast difference between the creatures that lived in the Cretaceous and those that lived in the Tertiary implied that the missing interval between them, which today we call the K–T boundary, represented as much time as all that has passed since. Today we know that time to amount to 65 million years, and the K–T boundary clay to represent to only a few thousand years at most.

To sum up, one type of uniformitarianism amounts to the statement that geology is a science; the second, which requires the adoption and maintenance of an a priori position regardless of the evidence, amounts to the statement that geology is not a science. Both cannot be true. But how then are we to account for the persistence of both in geological thought for nearly two centuries?

- The two types were so inextricably entangled that few students of geology ever realized that they were accepting "two-for-one." Since methodological uniformitarianism worked, the substantive variety tended to be accepted without anyone realizing that a fast one had been pulled.

- All catastrophism became equated with biblical catastrophism, to geologists an outmoded and shunned belief espoused only by scientific heretics.

- We are always attracted by a hero, and Lyell's writings had turned Hutton into the founder of geology. By espousing uniformitarianism, one stood tall beside the founding fathers.

- As uniformitarianism became dogma, it was re-espoused in each new geology textbook from Lyell to the present. Each generation of geologists learned uniformitarianism at its parent's knee, so to speak. Uniformitarianism had been around for so long that it never occurred to anyone to question it. (An exception who made public his doubts in his first scientific paper, written at age 25, is Stephen Jay Gould.[26])

- To call upon catastrophe to solve difficult problems diminishes the skills of generations of intelligent, hardworking geologists. It is too easy—a cop-out.

- Finally, only after World War II did the most dramatic evidence opposing uniformitarianism—the scarred and magnetized seafloor, which supported the notion of drifting continents (or moving plates), and the record of impact on other bodies in the solar system—become known.

We can now understand how the Alvarez theory ran into trouble on two grounds. First, it was catastrophic and contradicted the venerable doctrine of uniformitarianism. Second (and worse), it appealed to an extraterrestrial process, seeming to belittle the hard-won scientific achievements of generations of earthward-directed geologists. Add to these two reasons the natural resistance met by new theories and we have gone a long way toward understanding why geologists were far from delighted with the new Alvarez theory.

AN EXERCISE IN NEWSPEAK

Though modern geologists rejected a strict interpretation of Lyell's uniformity of state, by the 1950s most of those in North America had come to believe that at least the outer appearance of the earth, with its continents and ocean basins, had not changed dramatically—certainly continents had not drifted. The notion that seafloors spread out to plunge beneath continents, that the ocean basins are geologically young, that the continents have never been in the same place twice—all proved hard for those raised on uniformitarianism to accept.

After a generation to get used to plate tectonics, geologists have incorporated it into uniformitarianism. Because we can measure with laser beams, satellites, and global positioning systems the almost imperceptible movement of continents and the spreading of the seafloors, and can use the data to project backward to determine what the surface of the earth used to look like, the present can still be said to be the key to the past. Indeed, the way in which plate tectonics shows how older crust is buried in the mantle and recycled into new crust is reminiscent of Hutton's endless cycles. But Hutton and Lyell would certainly have rejected continental drift as impossibly antiuniformitarian. In any event, to say that one can infer the past positions of continents from their present positions and measured rates of motion is to appeal only to methodological uniformitarianism, which, as we have seen, is only to say that geologists proceed scientifically.

But meteorite impact as a force on the earth takes us into a new realm. Since we have never observed a large meteorite striking the earth, yet the existence of terrestrial craters tells us that they have, we cannot understand earth history by relying solely on processes that we can observe today. In short, the present is *not* a reliable key to the past. Just the opposite: To understand the role of impact cratering, we have to invert Hutton's aphorism and realize that, in the case of an event so rare as to fall outside human experience, the past must provide the key to understanding the present and the future.

By the time the Alvarez theory appeared in 1980, the space age had brought overwhelming evidence that impactors of every size had hit every object in the solar system countless times. We could of course stretch definitions to recognize the ubiquity of impact and claim that it amounts to a kind of uniformity, but this is equivalent to saying, "catastrophism is uniformitarian," an abominable oxymoron that would empty both words of meaning. As Ursula Marvin has pointedly said, "To regard the cataclysmic effects of impact as uniformitarian is an exercise in 'newspeak.'"[27]

Walter Alvarez drew the right conclusion about the proper place for uniformitarianism in geological thinking: "Perhaps it is time to recast uniformitarianism as merely a sort of corollary to Ockham's Razor, to the effect that if a set of geological data can be explained by common, gradual, well-known processes, that should be the explanation of choice, but that when the evidence strongly supports a more sudden, violent event, we will go where the evidence leads us."[28]

CHAPTER 3

STONES FROM THE SKY

Without help from a comet . . . I will give you a receipt for growing tree ferns at the pole, or if it suits me, pines at the equator; walruses under the line, and crocodiles in the arctic circle.[1]
Charles Lyell

In science as in life, timing is everything. A correct theory proposed before the time is ripe for its acceptance goes nowhere. The history of science is replete with theories ignored for years, decades, even centuries before their eventual acceptance. The most famous example is that of Aristarchus of Samos who anticipated by 18 centuries Copernicus's theory that the sun and not the earth is at the center of the solar system. In Aristarchus's day, however, Earth-centered astronomy did a good enough job of explaining the then rudimentary knowledge of the solar system, so that Aristarchus's theory was not "required." In 1866, the monk Gregor Mendel published his work on the laws of genetics in the proceedings of a local society of naturalists, but no one took notice. In 1900, 16 years after his death, Mendel's results were rediscovered. Continental drift had to wait half a century from Alfred Wegener's initial formulation in 1915 to the plate tectonics revolution of the 1960s and 1970s.

Why do ideas that eventually prove worthy often have to wait? Typically it is because they go against the grain of the current paradigm, leaving other scientists with no way even to think about them. When first proposed, they are often little more than inspired guesses with no supporting evidence. (Mendel was an exception; he had the evidence but published it where no one saw it.) The apparatus and techniques that will eventually provide experimental support often have yet to be invented. For example, only a few years prior to 1980, the Gubbio iridium anomaly could not have been detected, even if someone had been looking for it, because none of the available instruments were sensitive enough to detect iridium at the parts per trillion level.

The idea that a giant impact could cause mass extinctions, though consistently rejected by geologists, has a surprisingly long history, dating back at least to 1742, when Frenchman Pierre-Louis Moreau de Maupertuis suggested that comets have struck the earth and caused extinction by changing the atmosphere and the oceans.[2] His countryman, astronomer Pierre-Simon Laplace, wrote in 1813 that a meteorite of great size striking the earth would produce a cataclysm that would wipe out entire species.[3] In our own century, the distinguished paleontologist Otto Schindewolf sought an extraterrestrial cause for mass extinction. In 1970, Digby McLaren used his presidential address to the Paleontological Society to present the idea once again, leading some uniformitarians to assume that he could only have been speaking tongue-in-cheek.[4] American Harold Urey, winner of the Nobel Prize in chemistry, proposed in 1973 in the widely read journal *Nature* that impact was responsible for mass extinctions and the periods of the geologic time scale on which they are based.[5] Urey, who had published a variety of important research papers, had developed enough of a reputation in the earth and planetary sciences to be taken seriously, yet still no one paid any attention. These suggestions were catastrophist, unorthodox, and without evidence or predictions; therefore, even when made by distinguished scientists in important journals, they languished.

By 1980, when the Alvarez theory appeared, conditions had begun to improve. Iridium at the parts per trillion level was not easily measured, but it could be done at several laboratories around the world. The space age was nearly two decades old and the surfaces of other heavenly bodies were known in great detail—the map of the moon was more complete and accurate (when the ocean basins were included) than any map of the earth. It was impossible not to notice that, whatever its effect on the earth, impact had scarred every other object in the inner solar system innumerable times.

CRATERING IN THE SOLAR SYSTEM

The first person to observe lunar craters was Galileo. In 1609 he trained his telescope on the moon and saw the seas (the *maria*), the highlands, and some circular spots. He observed that as the terminator—the sharp line separating the light and dark sides of the moon—moved across, the far edges of the circles lit up before the centers. This told him that the rims of the circles were higher than their centers, which meant that they were depressions, or craters. From the length of their shadows, Galileo calculated the heights of the crater walls.

As the techniques of astronomy improved, the full extent of lunar cratering emerged and needed to be explained. Many eminent scientists and philosophers, including Robert Hooke, Immanuel Kant, and William Herschel (polymaths all), took a crack at the question. Almost to a person they concluded that the craters were the remnants of lunar volcanoes. The idea that impact created the craters was proposed from time to time but never taken seriously.

In 1892, the great American geologist G. K. Gilbert, whose interest in craters was fostered by his research on one in northern Arizona, began to study lunar craters by telescope. He observed that their shape, and their central peaks and collapsed terraces, showed them to be markedly different from terrestrial volcanic craters. For that reason, he concluded that they could not be volcanic but instead had to have been formed by impact. He conducted scale-model experiments and found that impact could indeed form craters, but that when the experimental projectile struck at an angle, the resulting crater was elliptical. Since meteorites arriving randomly on the surface of the moon surely must strike at an angle most of the time, at least some lunar craters should be elliptical, but as far as Gilbert could determine, all were circular. To reconcile experiment with observation, Gilbert proposed that a ring of solid objects in orbit around the moon, like the rings of Saturn, gradually released chunks that fell vertically onto the moon's surface. But since no evidence supported this theory, it too sparked little interest.

Gilbert conducted his tests in a hotel room, which meant that the experimental impactors fell at low velocities. He had no way of knowing that a projectile arriving at interstellar speeds is destroyed and its energy converted to an explosion that leaves a circular crater almost regardless of the angle of incidence. That knowledge had to await the early twentieth century and additional observations, many of them on the bomb craters that soon became all too available in the pockmarked fields of Europe.

Gilbert's negative conclusion essentially shut down research on the origin of terrestrial craters until, in 1961, a new era began when President John F. Kennedy, calling space the new ocean to be explored, declared that within that decade the United States would send a man to the moon and return him safely. Wisely, before astronauts were sent, unmanned vessels such as *Ranger*, *Surveyor*, and *Orbiter* mapped the moon. They found it densely cratered on every scale from thousands of miles to fractions of an inch. Missions from the Soviet Union and the United States showed that the lunar farside was also heavily cratered. Craters of every size saturated many lunar terrains, leaving no room for a new one without obliterating one or more previously existing craters. The larger craters had features not displayed by volcanic craters: central peaks, terraced rims, and rays of splashed debris. As this kind of evidence accumulated, it gradually became clear that lunar craters were not volcanic but were formed by impact. There was no scientific reason to believe that the other inner planets would have had a different history.

Astronauts brought back from the lunar highlands samples of impact breccia, rocks composed of broken, angular fragments embedded in a fine-grained matrix. This is what would be expected when impact breaks apart the rocks at ground zero, which are later cemented back together. The first Apollo astronauts, however, stepped out not on the highlands but onto a plain of basaltic lava, the type extruded by the Hawaiian volcanoes, showing that although there were no volcanic cones or craters any longer visible, at some earlier time vast sheets of lava had flowed out onto the lunar surface. When the lunar samples were dated, those from the highlands gave ages of 4.5 billion to 4.6 billion years, the same as the oldest meteorites and the calculated age of the earth, thus supporting the view that all the objects in the solar system have the same original age of formation. The volcanic rocks, however, which appeared both to earthbound geologists and astronauts to be the youngest lunar material, gave ages ranging from 3.1 billion to 3.7 billion years. In other words, the youngest moon rocks were almost as old as the oldest rocks on the earth, which date to 3.8 billion years. Geologists quickly realized that the moon had not had a continuous, steady geologic history like the earth—everything had been crammed into the first 1.5 billion years or so, after which the only significant process was meteorite impact.

The Apollo missions and the analysis of the returned samples had the ironic effect of debunking all the existing theories of the origin of the moon. As the full extent of cratering in the solar system came to be appreciated, Donald Davis and the versatile William Hartmann—scientist, artist, and author—proposed that early in the history of the solar system, a protoplanet the size of Mars struck the earth and blasted both itself and a large chunk of the earth into near-earth orbit, where the debris gradually amalgamated into the moon.[6] The mass that stuck became part of the earth's mantle. This has now become the theory of choice among planetologists. (In yet another example of a theory ahead of its time, Harvard's Reginald Daly proposed this very idea in the 1940s, but no one paid it any mind.[7])

The space age had hardly begun when it brought evidence of cratering on other planets. The *Mariner 10* spacecraft missions to Mercury in 1974 and 1975 found a surface as densely packed with craters as that of the moon. On Venus, *Magellan* found huge active volcanoes, vast lava flows, and a surface pocked with medium- to large-size craters. (Smaller craters were not found, probably because Venus's dense atmosphere causes small meteorites to burn up before they hit.) The *Voyager* mission showed that Jupiter's moons—Callisto, nearly as big as Mercury, and Ganymede, even bigger—are cratered on a lunar scale. Mars is not only heavily cratered but contains titanic volcanoes. And thousands of pockmarked asteroids float in space, some of them in orbits that cross that of Earth. Indeed, of all the bodies observed since the space age began, only three—Earth, and Jupiter's moons Io and Europa, all of them with active surfaces—lack obvious and plentiful craters.

Thus over the course of the twentieth century, impact cratering has gone from being viewed as extremely rare to being regarded as a paramount process in the history of the solar system. To have existed as a solid body in the solar system is to have been massively bombarded since the beginning. But where, then, are the craters that must have formed on Earth?

WHERE HAVE ALL THE CRATERS GONE?

After all, Earth not only has a much bigger cross-section than the moon to present to incoming meteorites, it also has a greater mass and therefore exerts a greater gravitational pull. Calculations combining area and mass show that at least 20 times as many meteorites should have hit Earth as hit the moon. The moon has 35 impact basins larger than 300 km in diameter, most of them nearly 4 billion years old. In the same period in its early history, 700 giant basins should then have formed on Earth. Such saturation bombing would have caused the entire surfaces of the moon and Earth to melt, forming giant magma lakes that persisted for millions of years (evidence for this is more visible on the moon than on Earth, where no trace remains of the early bombardment). Subsequently, both the moon and Earth were struck countless times, though not as often as during the first few hundred million years. Why then could the Alvarezes not find ready support for their theory in an abundance of terrestrial craters?

The first footprint at Tranquillity Base will outlast the pyramids and the tallest skyscraper—the moon contains no wind or water to erode the evidence of human visitation. Eternal, this giant fossil of the early solar system awaits our return. Earth, on the other hand, has been internally active since its creation, constantly renewing and reworking its surface materials, altering them beyond recognition. The erosive action of wind, ice, and water, and the large-scale effects of volcanic, plate tectonic, and mountain-building activity, have transformed the surface of Earth, and thus would likely have obscured or obliterated most terrestrial craters. The oceans cover 70 percent of Earth's surface and any crater that formed under the sea would be hidden. To add insult to injury, plate tectonics constantly recycles the seafloors, none of which therefore is older than about 125 million years. Since the end of the Cretaceous 65 million years ago, 20 percent of the seafloor has been carried down the deep-sea trenches that abut most continents, taking any accompanying craters with it.

It might be tempting to posit that terrestrial craters are rare because Earth, unlike the moon and Mars (but like Venus), has an atmosphere that incinerates incoming meteorites. We know that shooting stars suffer such a fate, but they come from tiny meteorites that weigh only a few grams and burn up completely. Slightly larger meteorites survive the trip but are slowed by atmospheric drag so that they arrive intact at nonexplosive velocities. But any meteorite larger than 40 m or 50 m in diameter blasts its way straight through the atmosphere and reaches the surface of Earth unimpeded and at cosmic velocities, whereupon it explodes, releasing tremendous amounts of energy.

It was not until the nineteenth century that scientists were prepared to believe that meteorites of any size actually came from space. Thomas Jefferson allegedly was told of the claim of two academics that they had observed a meteorite fall in Connecticut in 1807 and responded that he "would rather believe that two Yale professors would lie rather than that stones could fall from heaven."[8] (Even if apocryphal, this story does capture the sentiment of the day.) But actual meteor falls observed at Siena, Italy, in 1794 and at L'Aigle, France, in 1803, when more than 2,000 dropped, made it impossible to deny that stones do fall from the sky. Establishing the direct link between meteorites and impact craters was much longer in coming, however. It was first found at the end of the nineteenth century, near the Painted Desert, east of Flagstaff, Arizona.

METEOR CRATER

As one drives across the desert of northern Arizona, suddenly, and for no apparent reason, there looms ahead a mile-wide, nearly circular hole in the ground called Meteor Crater (Figure 4). In 1891, G. K. Gilbert, chief geologist of the U.S. Geological Survey and one of the most prestigious geologists in the world, attended a lecture in which this feature, known then for its raised rim as Coon Mountain, was described. Scattered around it were curious metallic fragments that were unlike any terrestrial rock and closely resembled iron

FIGURE 4 Meteor Crater, Arizona. [Photo courtesy of David Roddy and the U.S. Geological Survey.]

meteorites. Gilbert reasoned that if impact created this circular crater, the meteorite must have fallen vertically and could still be buried directly beneath the crater, where its iron magnetism would give it away. Surrounding the crater should be a mixture of ejected rock and meteorite fragments that together would have a greater volume than the now-vacated crater.

Announcing that he was "going to hunt a star," Gilbert and his assistants set out in October 1891 to measure the expected magnetism of the crater floor, but found none. The volume of ejecta turned out (by coincidence, we now know) to just match the volume of the crater. As a responsible scientist who followed where the evidence led, Gilbert had to conclude that impact had not created the crater. Such a conclusion was especially obligatory in this case, since Gilbert had set out his intended investigation of Coon Mountain as a model of the scientific method. He published his findings in 1896, four years after his hotel room experiments.[9] Having failed to find the predicted evidence of impact, Gilbert was forced to conclude that something other than impact, most likely a deep-seated gas explosion, had created the crater. Thus developed one of the great ironies in the history of geology: Gilbert correctly concluded that impact created the lunar craters, but incorrectly concluded that it had not created the most visible of all terrestrial craters. For four decades, Gilbert's enormous prestige and apparently meticulous methods put the theory of impact craters to rest.

The crater attracted not only scientific but commercial interest. Geologist and mining entrepreneur D. M. Barringer, "unaware that such ideas were geological heresy," as Marvin puts it,[10] decided that the meteoritic fragments at Coon Mountain meant that a large and valuable mass of meteoritic iron lay buried beneath the crater floor. He began his investigation in 1902 and continued it for 27 years, staking a claim and forming a company to mine the iron ore. Barringer sank exploratory shafts but found only the same fragments of meteoritic iron that had always turned up at Meteor Crater, as the cavity had come to be called. In 1929 he asked astronomer F. R. Moulton to calculate the amount of iron that should have been left behind. By this time, impact science had advanced enough for Moulton to conclude that the impactor would not have buried itself into the ground, it would have exploded, a fact of which no one in Gilbert's day was aware. Furthermore, Moulton calculated the mass of the meteorite at a mere 300,000 tons, far below Barringer's original estimate. Barringer's role in the search for impact products came to a tragic end only a few months later, just weeks after the stock market crash in 1929, when he died of heart failure. Though only fragments of meteoritic iron ever turned up at Meteor Crater, the work of Barringer and Moulton created an important legacy—the knowledge that at least one terrestrial hole in the ground was formed by an impacting meteorite.

Not long before Barringer died, Eugene Merle Shoemaker was born. He was just slightly too young to serve in World War II.[11] A young man in a hurry, he rushed through Los Angeles's Fairfax High School and the California Institute of Technology, emerging in 1948, at age 20, with bachelor's and master's degrees (and having been a school cheerleader along the way). He immediately joined the U.S. Geological Survey and went to look for uranium ore on the Colorado Plateau. Shoemaker recalled that one day in his first year with the U.S. Geological Survey, on his way to breakfast, it dawned on him that humans were going to "explore space." He thought, "I want to be part of it! The moon is made of rock, so geologists are the logical ones to go there—me, for example."[12] Shoemaker was right—humans were going to the moon, but unfortunately, medical problems prevented him from achieving his lifelong goal of being one of them.

By the mid-1950s, Shoemaker, always to be found where the cutting-edge geology was being done, was mapping nuclear bomb craters at the Nevada test site. His work on the Colorado Plateau had drawn to his attention a large cavity there that did not have a nuclear origin: Meteor Crater. Having satisfied the descendants of Daniel Barringer that he was not a disciple of Gilbert, in 1957 Shoemaker began the modern study of Meteor Crater. He used the time-honored methods of the field geologist: Study each rock unit close up and plot its position to produce a geologic map, the universal medium by which geologists communicate.

Sedimentary rocks of the type that rim Meteor Crater are deposited, naturally, with younger rocks above resting on older ones below. Yet at Meteor Crater, Shoemaker found just the opposite: The rocks on the crater rim were actually upside down geologically, with younger underneath older. He concluded that they had been blasted into the air, flipped over, and then had fallen to the earth again, but still upside down, forming a kind of upside-down layer cake. To lift huge masses of rocks and turn them over would have taken a great deal of energy. He too found the crater floor filled with breccia. Comparison with craters produced by nuclear test explosions allowed him to calculate that Meteor Crater had been formed by an iron meteorite weighing 60,000 tons, measuring 25 m in diameter, and traveling at 15 km/sec. Shoemaker calculated that the explosion was equivalent to the detonation of a 1.7 megaton nuclear device (85 times the magnitude of the bomb dropped on Hiroshima; recent estimates are higher) and destroyed all of the impactor save a few fragments. In 1964, the old generation and the new came together when Shoemaker guided Walter Bucher on a field trip to Meteor Crater. The evidence apparently convinced Bucher that the crater after all was due to impact, but he died before he could make his change of heart known.[13]

CRYPTOEXPLOSION STRUCTURES AND IMPACT MARKERS

The Steinheim Basin in Germany was one of the first cryptovolcanic structures to be described. Although it was initially put down to meteorite impact, this idea quickly gave way to the more orthodox view that the basin and others like it had been formed by ascending volcanic gases that fractured the rocks but whose associated lavas remained hidden, giving rise to the name *cryptovolcanic* for the structures (*cryptoexplosion* later became the preferred term). Curiously, however, the deeper the structures were probed, the less the rocks are deformed. If produced by volcanic activity, it should have been just the opposite. Some observant geologists wondered if the cryptoexplosion structures had been hit, not from below, but from above, and set out to find evidence.

Unfortunately, terrestrial craters, especially the older ones, are often so heavily eroded that only the barest trace of a circular structure remains, allowing them to be interpreted as either of cryptoexplosion origin or of impact origin, if not formed by some entirely different process. What was needed to resolve the issue was a marker, or set of markers, produced only by impact. It seemed theoretically possible that such markers exist, for the shock of impact is so intense and sudden that it produces conditions radically different from the low pressures typical at the earth's surface.

The first markers were discovered just after the turn of the century in a hill at the center of the Steinheim structure. The striated and broken cones of rock found there, known as shatter cones, had clearly formed from shock pressure, though its source was unknown. Some 40 years later, Robert Dietz, an early proponent of impact, studied the cryptoexplosion structure at Kentland, Illinois, located in the middle of the sedimentary rocks and cornfields of the American Midwest. In a large limestone quarry, he found shatter cones 6 feet long.[14]

The narrow ends of shatter cones tend to point back toward the center of their structure, showing that the fracturing pressure had come from there. Dietz believed that shatter cones would only be found at impact craters. Experiments (with Shoemaker participating) in which a gas gun fired pellets into limestone at 18,000 mph produced tiny but perfect shatter cones. Eventually they turned up in scores of other structures, including Meteor Crater, and came to be regarded, as Dietz had proposed, as an indicator of impact.

As noted earlier, the Sudbury structure in Ontario is one of the world's largest nickel ore bodies and one of the most thoroughly studied geologic features in the world. Decades of traditional geological approaches, however, had by the early 1960s produced no satisfactory theory to explain its origin. In a way analogous to the proposal of the Alvarez team that something completely outside normal experience had destroyed the dinosaurs, Dietz came up with the notion that Sudbury was created by a process so rare that no one had even thought to invoke it. In 1964 he proposed that Sudbury was a giant impact structure and, in his first visit, found the predicted shatter cones (Figure 5).[15]

Dietz went even further by endorsing the suggestion made in 1946 by Harvard's Daly that impact had also created the ancient South African structural dome known as the Vredefort Ring. (An impact structure that is very old and highly eroded would have ceased to exist as a topographic feature. All that would be left would be the concentrically warped rocks that were present at ground zero, hence the name "ring.") Dietz predicted the presence of shatter cones and again, in his first visit to Vredefort, found them. But shatter cones notwithstanding, most geologists thought that by proposing that impact had created the classic and intensely studied Sudbury and Vredefort structures, Dietz had crossed the line into heresy. At Meteor Crater, little was at stake and the misguided pro-impactors could muse as they liked. Sudbury and Vredefort were another matter; at these famous sites, decades of study and reams of publications placed reputations and geological orthodoxy on the line.

FIGURE 5 Shatter cones from Sudbury, Ontario. [Photo courtesy of R. Grieve and Geological Survey of Canada.]

Today we know from experiments that shatter cones mark the lowest pressures of impact, in the range of 5 gigapascals to 10 gigapascals. (Named in honor of a seventeenth-century mathematician and physicist, Blaise Pascal, a gigapascal [gPa] equals 10,000 times the pressure of the earth's atmosphere at the surface.) At slightly higher pressures—10 gPa to 20 gPa—quartz and feldspar, the two most common minerals in the earth's crust, begin to fracture in the characteristic crisscrossing planes, a few millionths of an inch apart, that I had originally seen in K–T zircon on the cover of *Nature.*

When a mineral with a certain crystal structure is subjected to sufficient heat and pressure, its atoms rearrange themselves into a structure that better accommodates the new conditions. For example, at low temperatures and pressures, pure carbon exists in the sheetlike structures that we call the mineral graphite. At higher temperatures and pressures, and under certain other conditions, carbon changes into the interlocking, three-dimensional structure that we call diamond.

Laboratory experiments show that quartz has two mineral phases that appear at high pressure but low temperature: The first to form is coesite, followed at about 16 gPa by stishovite. Thus the presence of stishovite at the surface means that the pressure at that point once reached 16 gPa. As far as we know, only meteorite impact produces such pressures. Coesite and stishovite were first discovered in nature at Meteor Crater. At pressures above 60 gPa minerals melt entirely. When these melts cool and freeze, they do not re-form the original minerals but instead harden into glasses that resemble ordinary igneous rocks, which explains how those at Sudbury, for example, could have been mistakenly identified.

The final impact marker is less direct. Scattered around the globe from Australia, through southeastern Asia, eastern Europe, the western coast of Africa, to Georgia and Texas, are large swaths of ground strewn with small glassy globules called *tektites,* after the Greek word for melted. Tektites usually have no relationship to the rocks with which they occur, leaving their origin a mystery. Their rounded, streamlined shapes and wide distribution suggest that they have traveled through the atmosphere while molten. For decades a debate raged over whether tektites had been splashed by impact off the earth or off the moon, with Nobelist Urey arguing for a terrestrial origin and Dietz and Shoemaker for a lunar one. Recently some tektites have been linked to particular terrestrial impact craters, showing that at least these tektites come from impacts on the earth.

Crater Types

Gilbert, and other early observers of lunar craters through telescopes, could see that they were of two types: smaller, rounded, bowl-shaped depressions, and larger, more complex structures with central peaks and collapsed rim terraces. Shoemaker, in his study of Meteor Crater, discovered why. When an asteroid or comet traveling at interstellar speeds strikes the earth, two powerful shock waves are created. The first, the explosive wave, travels downward through the target rocks, pushing them down and out. The second, the release wave, moves in the opposite direction. The shock and release waves interact in a complex manner, melting, vaporizing, and ejecting the rocks at ground zero. Fractured rock and crater walls fall back into the crater and mix with melt to form a breccia.

If the impactor is less than a few hundred meters in diameter, a simple crater (Figure 6) like Meteor Crater is formed. Such craters range up to about 4 km in diameter. Larger craters are not just bigger versions of small ones, as Tycho (Figure 7) illustrates. Such large craters start out in the same way as simple ones, but the greater energy released by the larger (or faster) impactor causes the rocks at ground zero to rebound to form the central peak. The crater rim cannot hold and falls in on itself to form terraces.

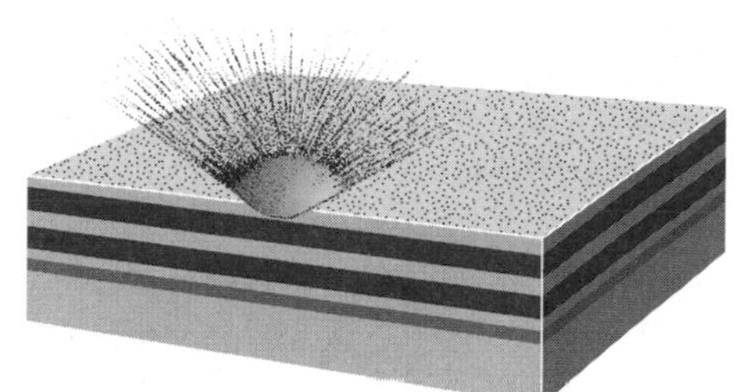
(a) contact

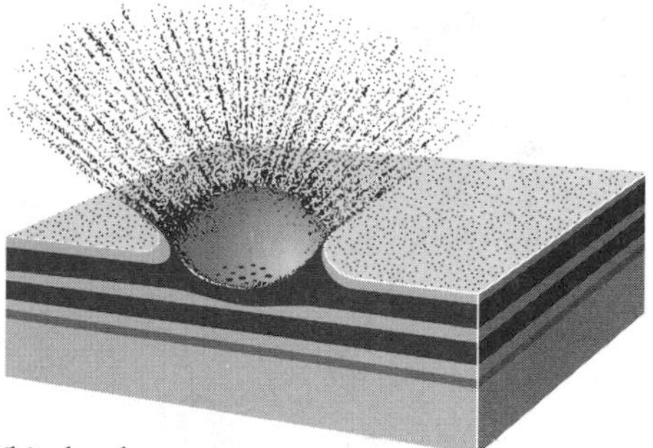
(b) shock waves

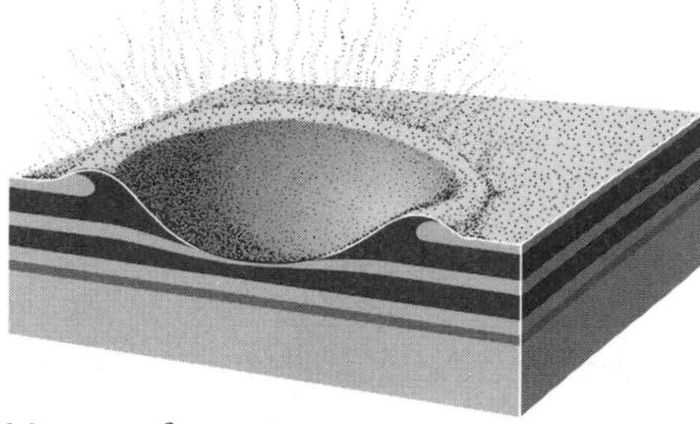
(c) crater formation

Figure 6 The formation of a simple impact crater. [After Don Gault.[16]]

Locations and Ages

As the photographs of other bodies taken from space began to be returned to the earth, the full extent of cratering in the solar system began to become apparent, at least to the more attentive, who concluded that there must be many undiscovered craters on earth, and set out to find them. They had their work cut out for them. A recent impact, like that at Meteor Crater some 50,000 years ago, leaves an obvious crater. But the earth is 4,500 million years old; most craters will have been so eroded that they no longer have any surface manifestation at all and may be detectable only through geological and geophysical methods. These techniques work because the shock of impact distorts the rocks at ground zero, raising central peaks and causing terraces to slump, as at Tycho. But these surface features, which on the earth are obscured by time and erosion, are underlain by structural ones—rock beds bent and twisted into concentric rings. Imagine, for example, that the moon had wind and water and that Tycho had been eroded for millions of years. Then the central peak, the terraces, indeed the crater itself would be gone, leaving no surficial hint that a crater had once been present at that spot. Buried in the rocks below the lunar soil, however, would be the bull's-eye imprint of the now vanished impact crater, detectable by geophysical

FIGURE 7 Tycho, a complex lunar crater 85 km in diameter. Note the central peak, the collapsed and terraced rim, and the hummocky ejecta deposits outside the crater. [Lunar Orbiter V-M125. Photo courtesy of the National Space Science Data Center, principal investigator L. J. Kosofsky.]

techniques. The magnetic, seismic, and gravitational properties of these rocks would reveal the bull's-eye pattern and show that it was once a crater.

By the time the Alvarez theory appeared, scientists had discovered some 100 terrestrial craters; today we recognize approximately 160, and the number increases by 3 or 4 each year. One-third are invisible at the surface, detectable only by geophysical properties. The first thing one notices about the distribution of terrestrial craters (Figure 8) is that although oceans cover 70 percent of the earth's surface, only a handful of impact sites have been found there (off Nova Scotia and eastern Russia). Since incoming meteorites would strike randomly, we assume that many more must have formed in the ocean basins, but that they have been hidden by younger oceanic sediments or carried down a descending tectonic plate. Most craters are located in the interiors of continents, as in North America, Europe, and Australia, which are geologically old, stable, and well studied (the older the surface, the more likely that it has been hit, but also the more likely that erosion will have removed the evidence). Approximately 20 percent of the craters are in Canada, a country that occupies only 1 percent of the total land surface of the earth. Oh, have the gods frowned on fair Canada? Emphatically not. Rather, the Geological Survey of Canada has mounted an intense search for impact craters in a country that contains a higher percentage of geologically old terrain than most. Few craters are known from South America and Central Africa, where rain forests make the search more difficult. About 60 percent of the craters discovered so far are younger than 200 million years, a period that represents only about 4 percent of

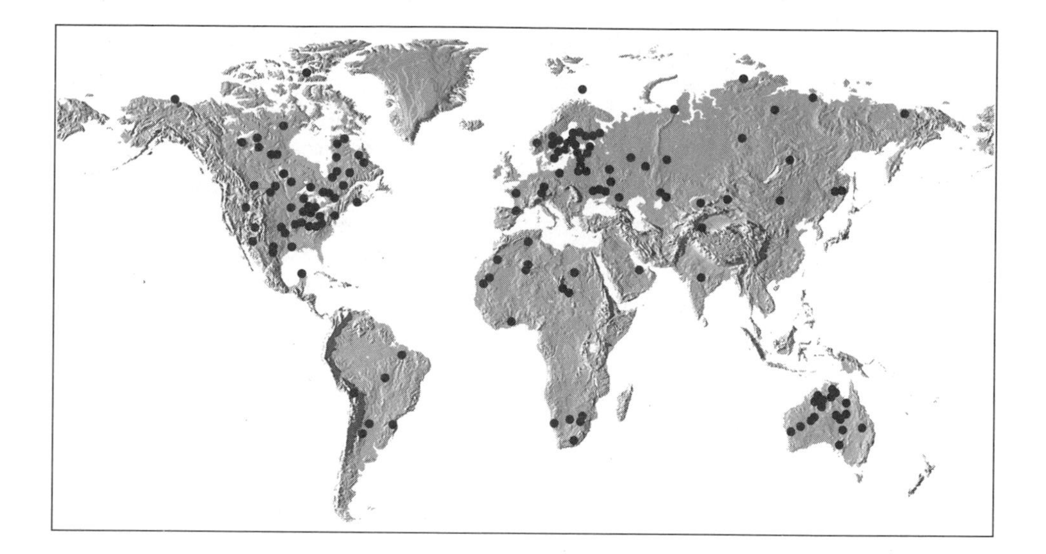

FIGURE 8 The distribution of known terrestrial craters. [R. Grieve, Geological Survey of Canada. Adapted from their web page at http://gdcinfo.agg.emr.ca/crater/world_craters.html.]

geologic time. Comparison with cratering on other bodies in the solar system shows that craters less than about 20 km in diameter are largely missing on the earth, presumably having been removed by erosion. Thus the observed record is biased toward younger, larger craters. More craters will be discovered on the earth in the future, but the smaller the original crater, and the older it is, the greater the likelihood that erosion has removed it forever.

FROM THE BACK OF THE MOON TO THE OUTBACK

What are the objects that strike the earth from space to form craters, and how is it that they can have such devastating effect, not only to dent the solid earth, but, as claimed by the Alvarezes, to play the starring role in dinosaur extinction? Astronomers have discovered that two types of cosmic objects are in orbits that sometimes intersect that of the earth: comets and asteroids.

Comets are "dirty snowballs"—mixtures of mineral dust and ices that evaporate under the heat of the sun to produce the visible tails that follow behind them for thousands of miles. The great comets of the 1990s, like Hayukatake and the spectacular Hale-Bopp, have been seen by hundreds of millions of people. Comets come from much farther away in the solar system than asteroids—from a vast cloud that surrounds the sun at an average distance about 40,000 times the earth-to-sun distance. Edmond Halley was the first to recognize that some comets are periodic, returning to our region of space with predictable regularity. Great comets had appeared over Europe in 1531, 1601, and 1682, and Halley figured out that these sightings were of one and the same comet. In another fine example of prediction, he claimed that the comet would reappear in 1758, and at definite intervals thereafter. Though he was not around to see it, Halley's Comet reappeared precisely on schedule and has continued to do so since, most recently in 1986. We can be certain that in 2061, this cosmic traveler will reappear, right on schedule.

In 1994, a rival to Halley for cometary fame appeared and, almost as suddenly, disappeared. Gene Shoemaker, his wife, Carolyn, and their colleague David Levy, had been searching the sky for comets and asteroids, carefully tracking the orbits of those they found, in order to determine whether the object might someday represent a threat to the earth. In 1993, through persistence and good luck, they spotted the comet that became known as Shoemaker–Levy 9 (SL–9; the ninth the trio had found together). Shortly after its discovery, SL–9 broke into pieces to present the image of a "string of pearls" to those who viewed it through telescopes. When astronomers plotted the comet's path, they realized to their surprise and delight that in July 1994 it was going to crash into Jupiter. It did so right on schedule, making planetary impact a reality seen by millions. Several of the fragments left dark spots on Jupiter that were larger than the earth. How fitting that Shoemaker, after a lifetime of studying craters, was not only the discoverer of the comet that was to produce the first planetary impact ever seen by human eyes, but was able to witness it.[17]

Asteroids are made of stone or iron; many are in orbits that cross that of the earth, meaning that a collision with our planet is theoretically possible. In March 1989, a previously undetected asteroid passed only 690,000 km from the earth, less than twice the distance to the moon. Calculations show that most earth-crossing asteroids cannot have been in their present orbits since the beginning of the solar system, or they would long since have collided with earth or been ejected into other regions of space. Some as yet unknown process must channel them into our region.

Because over recorded human history no recognizable impact crater has been formed (nor has impact cost a single life), it is reasonable to ask why scientists are confident that impacting asteroids and comets have caused great damage on the earth. The first point,

TABLE 1

Number of Craters of Different Diameters Produced During Last 100 Million Years, Based on Astronomical Observations.

Crater diameter	>20km	>30km	>50km	>60km	>100km	>150km
Asteroid source	190	58	8	3.2	0.3	0
Comet source	60	24	8	5	1.6	1
Total	250	82	16	8	2	1

noted in Chapter 1, is that by observing comets and asteroids astronomers can calculate how often one of a certain size is apt to hit the earth. Note from Table 1, from the work of Shoemaker and his colleagues, that the larger the crater, the more likely that it was formed by a comet. Astronomers and crater scientists usually express the frequency of crater formation as the number of craters larger than 20 km that form during a 100-million-year period on each 10 million km^2 of earth surface. Based on his astronomical observations, Shoemaker estimates the rate (for asteroids and comets combined) at 4.9 ± 2.9. In other words, Shoemaker reckons that during the last 100 million years, for each 10 million km^2 of earth surface (the earth has a total surface area of 500 km^2), 4.9 ± 2.9 (or roughly between 2 and 8) craters larger than 20 km have formed. From his observations of terrestrial craters, Richard Grieve of the Geological Survey of Canada obtains a rate of 5.5 ± 2.7, basically the same number as Shoemaker's. In recent years, Shoemaker and his wife Carolyn, when not searching for comets, were apt to be found camped in the Australian outback, mapping ancient impact craters. It was on such a field trip in July 1997 that he lost his life in a tragic automobile accident. Based on his most recent mapping and age dating, Shoemaker's last estimate of the frequency of impact for the old Australian craters, presented at a conference at the Geological Society of London in February 1997, was 3.8 ± 1.9. For impacts on the moon, including the lunar farside, it is 3.7 ± 0.4. It is simply astounding that all these rates, measured from the back of the moon to the outback, give the same answer.

Reason requires that we acknowledge that meteorite impact has been an inexorable fact of life in our solar system and on our planet. But even if we admit that, it does not tell us why meteorites are so terribly destructive. How can a ball of ice—even one much larger than the snowballs we hurled during our childhood with little effect—create a crater as large as those excavated at sites of nuclear explosions? The answer comes from a fundamental law of physics: Kinetic energy is equal to one-half the mass of an object times the square of its velocity. An object of little mass, if traveling fast enough, can contain a vast amount of energy. Comets and asteroids move at speeds far beyond our common experience—at cosmic velocities in the range of 15 km/sec to 70 km/sec, or 33,500 mph to 157,000 mph. On the average, comets travel at two to three times the speed of asteroids. Even though ice is less dense than rock or metal, because energy varies with the square of velocity and because of their greater speed, comets can do just as much damage as asteroids.

When a comet or an asteroid strikes the solid earth, the energy inherent in its great speed is converted into an enormous shock wave. An impact that creates a crater 10 km wide releases about 10^{25} ergs of energy (10^{25} = 10 followed by 25 zeros; the erg is a standard unit of energy; to lift a pound weight one foot requires 1.35×10^{7} ergs). The impact that produces a 50-km crater releases about 10^{28} ergs. For comparison, the 1980 eruption of Mount St. Helens released 6×10^{23} ergs, the 1906 San Francisco earthquake about 10^{24} ergs. The energy budget of the entire earth for one year—from internal heat flow, volcanic activity, and earthquakes—is about 10^{28} ergs. The asteroid envisioned by the Alvarezes would have released almost 10^{31} ergs. Because of their great speeds and the inexorable laws of physics, the impact of comets and asteroids releases more energy than any earthly process, placing impact in a destructive class by itself.

During the 1960s and 1970s, impact cratering came to be seen as a ubiquitous process in our solar system, one in which every solid object has been struck countless times by impactors of all sizes. Some of the projectiles were themselves the size of planets. As knowledge of the scale, frequency, and ubiquity of impact began to spread through the community of geologists, the notion that one might have occurred at the end of the Cretaceous, and even that it might have caused a mass extinction, no longer seemed quite so heretical. The saving grace of the Alvarez theory, and the reason it has proven so useful is that, in contrast to many other explanations of mass extinction, it can be tested. If the Alvarezes are wrong and the theory is false, the evidence would show it.

Part II

Was There a K–T Impact?

CHAPTER 4

THEORY ON TRIAL

It must be possible for an empirical scientific system to be refuted by experience.[1]
Karl Popper

OF PREDICTION AND PROOF

By the early 1980s, the importance of impact in the solar system was established as a fact, as was the presence of high iridium concentrations in at least a few K–T boundary clay sites. That the Cretaceous had ended with a great mass extinction was also a fact, though the suddenness of that extinction was disputed. The Alvarezes invented a theory that tied these facts together. To explain the observational facts is merely the first obligation of a theory; often several do a good job of explaining at least some of the observations. The theories that prove to have lasting value go further: They predict new facts that have yet to be discovered. If these predicted facts are subsequently found, the theory gains strength. Curiously, however, a theory is never completely proven. The possibility always exists that some new evidence will come to light to discredit the theory, or that some clever scientist will come up with an alternative theory that explains more of the facts. Luis Alvarez never went so far as to claim that the meteorite impact theory had been proven, though he came perilously close. Typically he would assert only that the theory had met a large number of its predictions (and "postdictions," which are reasonable predictions that happened not to be thought of until later). Being human, however, he was not above, in Tennyson's phrase, "believing where we cannot prove."[2]

German philosopher Karl Popper has done more than anyone to advance the notion that scientific theories may be disproven, but never proven. He argued that for a theory to be called scientific, it must be possible to disprove, or falsify, the theory.[3] For a theory to qualify as part of science, it must be possible to devise tests that, if a theory is wrong, will reveal it as wrong. If no such tests can be devised, then the theory is not useful, at least for the time being. This is one reason why premature theories languish: No one can think of anything useful to do with them. Popper did allow that theories could be "corroborated"; that is, they could prove their mettle by standing up to a succession of severe tests. Corroborate is a good word—it means "to strengthen or support with other evidence; make more certain." Corroboration falls short of proof, but shows that research is heading in the right direction.

Though philosophers and historians of science debate the utility of Popper's formulation and are apt to go on doing so, it jibes with our common sense to say that science advances not by proving theories right but by weakening them until they are falsified. Looking back at the history of science, it is clear that this is the way it works. Yet if one were randomly to select a scientist at work and ask, "What are you doing?" one would be apt to get the answer: "I am confirming such and such a theory." In their daily lives, most scientists try to confirm or extend theories, not to falsify them. In part this is because scientists are rewarded for breakthroughs, not for falsification. Rewards aside, however, human beings will not spend long hours and entire careers searching for falsity. Thus a contradiction exists between the way individual scientists behave and the way science as a whole evolves—as the cumulative result of the work of all scientists. A host of them, each trying to shore up their favorite theories, will in time lead to the falsification of the weakest, to the great disappointment of its proponents but to the advancement of science overall.

ALVAREZ PREDICTIONS

The Alvarez theory revolves around two key hypotheses: (1) 65 million years ago, a meteorite struck the earth, and (2) the aftereffects of the impact caused the K–T mass extinction. Since one can accept the first without accepting the second, they need to be kept separate (although the Alvarezes did not). In the rest of this chapter, I will examine the evidence for the first half of the theory. (The second half is covered in Chapters 8, 9, and 10.) Although Luis Alvarez himself identified 15 pre- and postdictions, not all are of equal importance. I will focus on six predictions that if confirmed would be especially corroborative and that can be identified largely by using common sense. If several of the six predictions turn out to be false—certainly if all did—the Alvarez theory would have to be abandoned. On the other hand, if most or all are met, the theory would be strongly corroborated.

An explosion of research effort followed publication of the initial Alvarez paper as geologists around the world set to work, some seeking to confirm its predictions while others tried to refute them (in principle, intent does not matter as long as the rules are followed). Key events in the refinement of the theory were the conferences held at the Snowbird ski resort in Utah in 1981 and 1988, and in Houston in 1994.[4] The conferences brought together the leading workers in the new field of impact studies and a variety of other specialists, proponents of the theory and opponents alike, and provided a forum for papers and for debate that went on into the wee hours. For tracing the evolution of the Alvarez theory, the reports from these conferences are indispensable.

Here are six of the most important predictions made by the Alvarez theory, followed in each case by the corresponding findings.

PREDICTION 1: Impact effects will be seen worldwide at the K–T boundary.

A global catastrophe would leave global evidence. Most if not all K–T boundary sites around the world will contain an iridium anomaly, though the concentration might be greater at sites closer to the ground zero of meteorite impact. At some locations, however, subsequent geologic processes might have removed iridium or even eroded the boundary layer entirely away, leaving a gap in the rock record. Thus although the absence of iridium from a few K–T boundary clay sites might not falsify the Alvarez theory, were iridium found nowhere other than in Italy and Denmark, the theory would be in trouble.

FINDINGS

By the time of the first Snowbird Conference in 1981, only a year after the original paper in *Science*, the number of sites with confirmed iridium anomalies had risen to 36. By the end of 1983, it had reached 50; by 1990 it had climbed to 95; today it is well over 100. Iridium concentrations in the boundary clays are the highest ever measured in terrestrial materials. Only a few K–T sections lack iridium.

One site was of critical importance, for it was the first in which the rocks studied had been deposited not in seawater but in fresh. Some had claimed that impact of a meteorite was not the only way to get iridium into a rock layer. Seawater contains trace amounts of the element; perhaps there were processes that could somehow concentrate the iridium from a large reservoir of seawater into a particular rock layer. Iridium might be absorbed selectively on the surfaces of the clay minerals, for example. Or, perhaps the clay and iridium were once dispersed minutely throughout a thick, marine limestone bed that slowly dissolved away, leaving behind only the insoluble clay and iridium. These ideas might have applied to rocks deposited in the sea, but not to those laid down as sediments in freshwater, which contains even less iridium and where there is no opportunity to tap a vast reservoir. The discovery of a strong iridium anomaly in rocks from the Raton Basin in New Mexico and Colorado, rocks recognizable as having formed in freshwater, put the idea of seawater extraction to rest.[5] (Luis Alvarez, with the advantage of hindsight, said that the occurrence of the iridium spike in freshwater rocks should have been one of his predictions.) At the exact level of the Raton iridium spike, several Cretaceous pollen species went extinct and ferns—which are opportunistic and move in after other species disappear—proliferated.

PREDICTION 2: Elsewhere in the geologic column, iridium and other markers of impact will be rare.

If high iridium concentrations come from meteorites, they will not be found in most other rocks. If the indicators of shock described in Chapter 3—shatter cones, shocked quartz, coesite, stishovite, and tektites—are produced only by impact, they too will be rare to nonexistent in other geological settings.

(This is an appropriate place to note that the K–T mass extinction was one of many times during which substantial numbers of species disappeared. Paleontologists have identified five, including the K–T, that were especially severe. If impact is responsible for any others of the "Big Five," they too might show an iridium spike and impact markers. However, the presence or absence of indicators at those horizons would have no direct bearing on the Alvarez theory, which applies only to the K–T event. The possibility that impact might have caused more than one mass extinction is a related but separate theory that I will address later.)

FINDINGS

It is obviously impossible to search for iridium in every rock on the surface of the earth. Frank Kyte and John Wasson of UCLA did the next best thing by measuring iridium content in a long, continuous core of sediment pulled up from the deep seafloor in the Pacific.[6] It captured the sedimentary record from about 35 million years ago all the way back to the K–T boundary at 65 million years. They found iridium levels above background only at the K–T boundary. As far as we know, high iridium concentrations are exceedingly rare in terrestrial rocks.

PREDICTION 3: Iridium anomalies will be associated with proven meteorite impact craters.

The Alvarezes started with an iridium spike and inferred an impact; it should be possible to move in the other direction as well. That is, it should be possible to find a crater whose origin by impact is undisputed, predict where the corresponding iridium-enriched ejecta will be located, and go find it. But since it is hard to detect terrestrial craters in the first place, and since erosion will have removed some ejecta layers, the absence of such a connection would not falsify the Alvarez theory.

FINDINGS

Two craters have been found to have associated iridium-rich ejecta layers. One is the 600-million-year-old crater at Acraman, South Australia, whose ejecta deposit contains not only iridium but other platinum group metals as well as gold.[7] This ancient crater has been so deeply eroded that only a multiringed scar remains. Its ejecta, even though located more than 300 km away, can still be tied confidently back to the crater. The other is the 40-km diameter, 143-million-year-old Mjolnir crater, in the Barents Sea north of Scandinavia, which was detected through geophysical methods.[8] A diligent search led by a group of Norwegian geologists found its ejecta layer, which contained both iridium and shocked quartz, in a core taken 30 km from the crater's center.

At first it may seem surprising that it is so difficult to connect iridium-rich ejecta layers to their parent craters. But remember how difficult it is to recognize terrestrial impact craters, and to find the thin ejecta layers, in the first place. Comparing it to the search for a needle in a haystack may be optimistic. In any case, the two examples prove the principle. As the science of crater detection improves, other ejecta layers will be tied back to their parent craters.

PREDICTION 4: The boundary clay layer will generally be thin and of worldwide distribution.

The immediate effects of a giant impact take place in minutes or hours; the secondary ones may last for hundreds or at most a few thousand years. On a geologic time scale, even these are instantaneous. Thus the boundary layer will be thin everywhere except, perhaps, at sites closer to ground zero. The layer ought to be found globally, though erosion might on occasion have removed it. If a thin layer is found worldwide at the K–T boundary, it would be the first universal geologic marker—rock formations ordinarily are no more than regional.

FINDINGS

Around the world, the K–T boundary is marked by a thin clay layer, almost always with high iridium levels. (As we will see, in North America there are two boundary layers, with the thicker one on the bottom.) No other rock unit extends over even a single continent, much less over all of them and the seafloors in between. The very existence of this universal layer is evidence of a rare, perhaps unique, geologic event, and is as strongly corroborative a piece of evidence for the Alvarez theory as any.

PREDICTION 5: The K–T boundary clays will contain shock metamorphic effects.

Known markers of impact—shocked quartz grains; coesite or stishovite; glassy, tektitelike spherules—will be found in the boundary clays. The presence of these accepted indicators would provide much stronger corroboration to doubting geologists than the iridium spike, which prior to the Alvarez discovery was unrecognized as an impact marker.

FINDINGS

In 1981, geologist Bruce Bohor of the U.S. Geological Survey decided to look for shocked quartz at the K–T boundary and applied for a Survey fellowship (ironically named in honor of G. K. Gilbert). Turned down by the fellowship panel (which included a specialist in shocked quartz), Bohor reapplied, only to be rejected again. Showing admirable resolve, he went ahead on his own and shortly did locate shocked quartz at the K–T boundary in a 1 cm thick Montana claystone that also contained both a large iridium spike and a pollen extinction.[9] Bohor's discovery was crucial in making believers out of many geologists. First of all, one of their own, rather than a know-it-all physicist, had made the discovery. Second, instead of being based on an invisible element, shocked quartz was a tried-and-true indicator that geologists had discovered

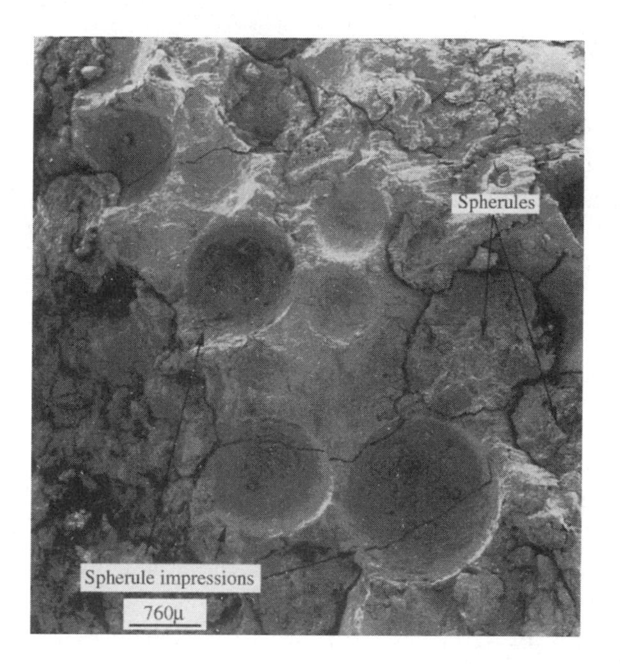

FIGURE 9 Imprints left by K–T spherules where they fell in soft clay. From a drill core that penetrated the K–T boundary underneath New Jersey. [Photo courtesy of Richard Olsson, Rutgers University.[10]]

themselves—it was "invented here," and could be seen with a microscope. Bohor and others went on to find shocked quartz at many other K–T boundary sites around the world. Stishovite, which provides evidence of extreme pressures, has been found at several.

Many K–T sites have yielded millimeter-sized spherules that look for all the world like microtektites externally but that internally are composed not of glass but of various crystallized minerals. Some show beautiful flowlines on their surfaces. They have been studied extensively and have a mineralogy unlike anything geologists have seen before. The pro-impactors interpret them as droplets melted by the shock of impact and blasted into the earth's atmosphere, where they solidified and fell to earth (Figure 9), subsequently recrystallizing into the minerals that we now find.

PREDICTION 6: A huge impact crater formed 65 million years ago. If it has not disappeared, it may yet be found.

If the Alvarez theory is correct, there once existed, and we can hope there still does exist, a huge crater exactly 65 million years old. Failure to find it would not falsify the theory, however, because the crater could easily have escaped detection. The meteorite might have hit somewhere in the two-thirds of the earth's surface that is now covered with water, leaving the crater hidden under younger sediments. It might have landed in the 20 percent of the seafloor that has subducted (carried down beneath an overriding tectonic plate) since K–T time. The crater might be buried under the polar ice caps. It might have struck on land but now be so eroded as to be undetectable, or it might be buried there beneath younger rocks. It might have triggered a volcanic eruption and now be covered with lava. Finding the crater thus would require more than good science—it would require good luck.

FINDINGS

Locating the K–T impact crater obviously would provide the most corroborative evidence of all, but prior to 1990, no crater of the right age and size had been discovered. Indeed, the only candidate much discussed was the buried structure at Manson, Iowa, but it seemed too small to create a worldwide catastrophe and, some critics said, was a cryptoexplosion structure formed by underground gas explosions, as Gilbert and Bucher had claimed for Meteor Crater. (Chapter 7 is devoted to the search for the K–T impact crater.)

The six predictions just reviewed are specific to the Alvarez theory, but there is a seventh prediction that can safely be made whenever a theory with far-reaching implications is explored.

PREDICTION 7: Unanticipated discoveries will be made.

As theories are explored, unexpected discoveries almost always turn up. Sometimes these surprising findings turn out to strengthen a theory; sometimes they provide critical evidence that helps to falsify it. With the advantage of 20-20 hindsight, one can often see that a particular discovery could have been anticipated and stated as a prediction.

FINDINGS

Prior to the Alvarez discovery, little was known about the clay layer, but now a host of techniques were applied to it. Three scientists from the University of Chicago made one of the most astonishing finds.[11] Searching the Danish K–T clay for a possible meteoritic noble gas component, they found large amounts of soot, which was missing in the other late Cretaceous rocks and marine sediments that they analyzed for comparison. If the clay layer had been deposited suddenly, for which there is much independent evidence, then such a large amount of soot could only have come from global wildfires in which possibly as much as 90 percent of the total mass of living matter on the earth burned. Supporters of the impact theory naturally found the presence of the soot highly corroborative. Opponents pointed out, however, that the conclusion depends on the assumption that the clay layer was deposited rapidly; if it were not, the levels of soot would not be extraordinary.

Soot was not the only unexpected substance. Amino acids, the building blocks of proteins, and ultimately of life, are ubiquitous on the earth but also occur in a class of meteorites called the carbonaceous chondrites. Canadian scientists reported that the boundary clay contains 18 amino acids not otherwise found on the earth.[12]

Osmium is a platinum metal almost as rare in crustal rocks as iridium. Karl Turekian, a geochemist at Yale, noted that the ratio of two isotopes of osmium, Os 187 and Os 186, in meteorites is approximately 1:1, but in rocks of the earth's crust it is higher than 10:1. Although chemical and geological processes concentrate some chemical elements and deplete others, the ratios of the isotopes of heavy elements such as osmium tend to remain constant. This resistance to alteration is best illustrated by the enormous effort required in the Manhattan Project to separate fissile U 235 from U 238, which is 100 times as abundant naturally. Even the heat and shock of meteorite impact would not change the ratio of isotopes as heavy as those of osmium, and therefore they can be used as a tracer and proxy to reveal the origin of the iridium in the K–T layer. If the Os 187:Os 186 ratio in the boundary clay turned out to be close to the meteorite ratio of 1:1, then the osmium would likely be extraterrestrial, as would the iridium, an almost identical element. If the osmium isotopic ratio were much higher, then both the osmium and the iridium would likely be of crustal origin, weakening the Alvarez theory. According to David Raup, at Snowbird I, Turekian "made it abundantly clear that he expected to find ordinary crustal isotope ratios and that his study would show that the impact theory was neither necessary nor credible."[13] A year and a half later, Turekian and a colleague reported that the osmium isotope ratios in the boundary clay were closer to meteoritic than to terrestrial levels.[14] The osmium test was less definitive than had been hoped, however, because an osmium ratio of approximately 1:1 turned out to mark not only meteorites but volcanic rocks from the earth's mantle. Thus a low osmium isotope ratio could indicate a mantle source as well as an extraterrestrial one. A recent study of samples from across the last 80 million years of earth history, however, turned up a low osmium isotopic ratio only at the K–T boundary, again strengthening the possibility that the osmium and iridium came from space.[15]

Spinel is a rare mineral that sometimes forms a variety of ruby. Many of the K–T clays contained a nickel-rich variety of spinel previously found only in material worn off of meteorites. Furthermore, the highest spinel abundances occurred at exactly the same place in K–T sections as the iridium spike—at some locations each gram of boundary rock contained more than 10,000 spinel spherules.

Until recently, diamonds occurred in nature only in rocks believed to have originally formed deep within the earth (where heat and pressure are high), and that subsequently were elevated to the surface. Within the last few decades they have been produced in explosion experiments and found in meteorites, where the diamonds are so tiny as to be barely detectable. The hope that diamonds would also show up at terrestrial impact sites led Canadian scientists David Carlisle and Dennis Braman to search the K–T boundary clay in Alberta, where they immediately found them.[16]

Now the story gets even more interesting.[17] As the Soviet Union began to collapse, reports started to emerge that scientists there had not only found diamonds at several of their impact sites, but in numbers reaching into the millions. The most thoroughly tested crater was the 35-million-year-old, 100-km-wide Popigai Crater in Northern Siberia, which the Soviets probed with over 500 boreholes. Most of the diamonds there were tiny, but some were as large as peas. (Although none are of gemstone quality, they may prove useful for industrial purposes.) A British team searched the Ries Crater and soon found diamonds by the billions in the melt rock, which had been the source of the stone for the town hall and the church in Nordlingen, the medieval German town located within the crater. The citizens of that town, unbeknownst to them, had been surrounded all their lives by innumerable diamonds formed 15 million years ago by a giant impact. The Ries diamonds occur in association with silicon carbide, like diamond a rare and hard mineral. From this association and various other chemical indicators, the British scientists concluded that the diamonds and silicon carbide had not formed directly as a result of shock but rather had crystallized in midair from the white-hot impact fireball. If they are correct, then diamonds should be found at other impact craters and provide an excellent marker of impact. The search was immediately extended, and by mid-1996 diamonds had been found in each of the eight impact sites studied. No diamonds occur in rocks immediately above and below the K–T boundary—only right in it. The finding of billions of diamonds at impact sites and K–T locations must rank as the most surprising and important of the unexpected discoveries triggered by the Alvarez theory.

THE FIRST HURDLE

How well, then, has the Alvarez theory done in meeting its first hurdle: being tested against the six predictions? It can be summed up as follows:

1. The K–T iridium anomaly is found worldwide.
2. With a few exceptions, the iridium enrichment is not found at other geological horizons.
3. Impact does produce distant ejecta deposits enriched in iridium.
4. Almost everywhere the K–T boundary itself can be located around the earth, the boundary clay layer is present. Except for a few sites the layer is thin.
5. Accepted indicators of impact—quartz with planar deformation features, coesite and stishovite, and spherules that resemble microtektites—are present at many K–T locations.
6. The impact crater may have been found (see Chapter 7).

In addition, the seventh prediction, of unexpected discoveries, has not only been met, the surprising findings of soot, amino acids, meteoritic osmium isotopes, spinel, and finally diamonds, some of which are difficult or impossible to explain by terrestrial causes, help to corroborate the theory.

A neutral observer examining this evidence would have to conclude that the Alvarezes had a strong initial case for the impact half of their theory. But when a theory has potentially revolutionary consequences, few observers are neutral. Opponents of the theory immediately began to attack in earnest, arguing that either the tests were not valid, or, if they were, had failed. In this view, the iridium anomaly is not restricted to the K–T horizon; indeed, opponents claimed, iridium is not a true marker of impact at all. The boundary clay shows no sign of a meteoritic component; besides, shocked quartz and spherules are not diagnostic indicators of impact. The critics claimed repeatedly to have falsified the theory or to have found an alternative that fulfilled the predictions at least as well and that was consistent with uniformitarian doctrine to boot. The Alvarez team rebutted, the critics countered, and thereby was produced one of the most bitter scientific rivalries since the great controversy between dinosaur hunters Edward Drinker Cope and Othniel Marsh, who in the late nineteenth century quarreled openly for two decades about their interpretation of the dinosaur evidence, with one even accusing the other of stealing his fossils.

Chapter 5

Counterattack

If you start to take Vienna—take Vienna.[1]
Napoleon Bonaparte

The first to offer a detailed attack on the Alvarez theory were not paleontologists, as might have been expected since they were apparently the most offended, but a geophysicist from Dartmouth College, Charles Officer (Figure 10), and his colleague, geologist Charles Drake. Officer had had a distinguished career as a seismologist in industry; Drake was one of the most respected American geologists, having been president of the American Geophysical Union from 1984 to 1986, at the height of the controversy, and also president of the Geological Society of America. They began their rebuttal with two papers in *Science*,[2] which together comprised a three-part plan of attack:

1. Falsify the impact theory by showing (a) that the K–T event took place at different times around the world (and therefore could not have been the result of an instantaneous global catastrophe), and (b) that the transition from Cretaceous to Tertiary fossil species was too gradual to be consistent with an instantaneous extinction. (I will call these arguments 1a and 1b, respectively.)
2. Show that the evidence of iridium anomalies, shocked minerals, and spherules was far from diagnostic of impact and often was not even consistent with it (argument 2).
3. Substitute for impact another process that explains the evidence at least as well and that does not rely on a nonuniformitarian deus ex machina (argument 3).

FIGURE 10 Professor Charles Officer. [Courtesy of H-O Photographers.]

PREEMPTIVE STRIKE

If the Alvarez theory is correct, the change from Cretaceous time to Tertiary time happened instantaneously, everywhere around the globe, in which case the K–T boundary clay can represent but a mere eyeblink of geologic time—a few hundred or a few thousand years at most. Furthermore, the boundary would have to be the same age everywhere. Officer and Drake claimed to have evidence that, on the contrary, the age of the boundary differs by hundreds of thousands of years at different locales. If the K–T boundary has one age at one site and quite a different age at another, it obviously was not created instantaneously and the impact theory is falsified. On the other hand, without impact as its cause, the K–T boundary is likely to have somewhat different ages at different locations around the globe. This surprising fact has its roots in the way geology began.

As their understanding of the earth developed, early geologists began to recognize that they could define distinctive rock units that differed both from the older rocks below and the younger ones above. More than a century ago, geologists began to give names to these characteristic units that they could trace over wide distances: Cambrian, Ordovician, Devonian, and so on. In this way the standard geologic column—the ideal sequence if all rock units were present, none having been removed by erosion—was constructed. It is the basis for the subdivisions shown in Figure 2. Since in a given geographical area only a limited portion of the geologic column is exposed, its fundamental units were sometimes situated in entirely different countries, leaving no way to correlate them precisely. Near the end of Cretaceous time, different types of rocks were being deposited in different environments: a limestone on the undersea shelf near one continent; a sandstone on a beach halfway around the globe; a shale in a swamp on another continent. In the absence of a worldwide, short, terminal event, these processes would not have ended at exactly the same time, and thus the K–T boundary would have a slightly different age at different places around the world.

Compare, for example, the K–T boundary at Gubbio with that at Hell Creek, Montana, source of *Tyrannosaurus rex* and the best-studied dinosaur fossils in the world. As shown in Figure 3, the K–T boundary at Gubbio is easy to spot—you can place your finger right on it. On the other hand, at Hell Creek the boundary is exceedingly difficult to locate, or even to define (it is described in the literature as "above the highest dinosaur fossil and just below the level of the lowest coal bed," neither of which occurs at Gubbio). Let us suppose, however, that we could find a boundary at Hell Creek that we believe demarcates the K–T. How could we determine whether it is of exactly the same age as the K–T boundary at Gubbio? We cannot do so by comparing fossils, because those at Gubbio are marine microfossils (foraminifera) whereas the rocks at Hell Creek were formed in freshwater and contain dinosaur and mammal bones, but no forams. One way to determine rock ages precisely is through the use of a pair of elements in which one, the parent, decays radioactively into the other, the daughter, as when uranium decays into lead. If one knows how much uranium is present in a sample, and how much lead, and one knows how fast uranium decays into lead (the half-life), one can calculate how long the process of decay has been going on in that sample, and thus derive the age of the rock. But none of the rocks from Gubbio has enough of the parent element or occurs in close association with the volcanic rocks that are best suited for parent-daughter (radiometric) dating. Even if the parent-daughter methods could be used, however, they are insufficiently precise for exact correlation.

Very well, if we cannot show that two rocks from the same section of the geologic column on different continents are exactly the same age, can we do the opposite and show that their ages differ measurably? Not by using forams on one continent and dinosaurs on another, nor by using radioactive parent-daughter ages, where the same lack of precision remains a limitation. But there is one possibility: the magnetic reversal time scale that led Walter Alvarez to Gubbio in the first place. It would be ironic if that same scale could be used to falsify the Alvarez theory, but that is exactly what Officer and Drake claimed to have done. Here is the basis for their approach.

As briefly described in Chapter 1, over the past few decades, geophysicists have established that the earth's magnetic field has repeatedly reversed its polarity.[3] The north magnetic pole has acted alternately in the way we define a north magnetic pole as acting, then as a south magnetic pole, then as a north pole again, and so on, over and over, throughout hundreds of million of years. These reversals have affected the entire magnetic field, all around the globe. During a period of reversed magnetism, a compass needle, which seeks a north magnetic pole, would instead point toward magnetic south. It is not known why the earth's magnetic field reverses, though supercomputer modeling is beginning to shed light on the mystery. But remember that we discover facts and invent theories. We have discovered that the earth's magnetic field has reversed itself hundreds of times, on the average about every 500,000 years; so far we have not been inventive enough to figure out why.

Magnetized rocks of different ages around the world have been dated using one of the radioactive parent-daughter pairs, and thus we know, within the precision of those methods, when each reversal occurred. The K–T section of the magnetic reversal time scale is shown in Figure 11. The major intervals are represented by shaded and light bands, called chrons, numbered and designated R for reversed and N for normal. Ordinarily, the magnetic time scale suffers from the same lack of precision as the other methods and cannot be used to show that two rocks have precisely the same age. For example, if all we know is that two rocks belong to Chron 29R, we have not pinned their ages down to better than ±750,000 years, the duration of that chron. But suppose on the other hand that we can establish that one rock unit belongs to 29R while the other belongs to 28R. We can then be certain that the two are *not* of the same age and that their ages must differ by at least 800,000 years, the duration of intermediate Chron 29N. (The absence of 29N in the region under study means either that rocks from that age were never deposited, or that they were subsequently removed by erosion.) Thus, paleomagnetism may be able to show that although two rock layers date to the same general part of the geologic time scale, they do not have identical ages. This was the opening that Officer and Drake hoped to exploit.

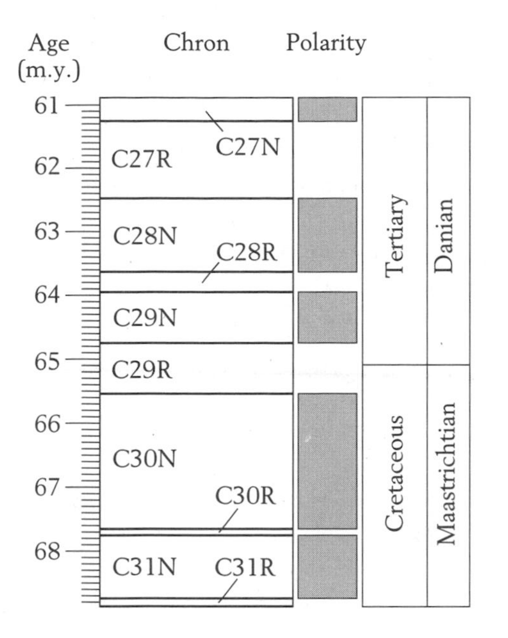

FIGURE 11 The magnetic reversal time scale around K–T time. Dark bands represent periods of normal magnetization (like today); light bands show reversed magnetization.[After Berggren et al.[4]]

In their 1983 paper in *Science,* Officer and Drake reviewed the literature on magnetic reversals near the K–T boundary, focusing their attention on six deep-sea drilling sites and nine continental sites, including Gubbio and Hell Creek. Most fell into Chron 29R, providing no support for their thesis, but three did not, giving Officer and Drake a foothold. A sample that came from a deep-sea core, and another that came from the San Juan Basin in Colorado, appeared to belong to 29N; the third sample, from Hell Creek, they assigned to 28R. If these interpretations were correct, the K–T boundary was not the same age everywhere and therefore could not have been produced by an instantaneous event. The Alvarez theory would have been preemptively falsified, the battle would have been over before it began, and geologists could return to business as usual.

Officer and Drake's argument received a quick rebuttal from the Alvarezes, who accused the pair of breaching scientific etiquette by failing to cite any of the papers presented at the 1981 Snowbird I conference, even though first the abstracts and then the entire volume of papers from the conference had been published and even though Drake himself had not only attended the meeting but, months before the *Science* paper appeared, had published a critique of some of the papers presented there.[5,6] Why was this a serious error? Because scholarship is cumulative, with each generation standing on the shoulders of those who have gone before. To fail to cite relevant papers that one knows about is a grievous error (to overlook one you should have known about is bad enough): It cheats the authors of those papers of their rightful recognition; it misleads readers who are not expert in the subject at hand; and it avoids contradictory evidence, thereby falsely fortifying your own position. Most scientists would say there are only two reasons for failure to cite a relevant paper: ignorance or dishonesty. Neither seemed to be an explanation in this case, first because Officer and Drake clearly knew of the Snowbird I conference and report, and second because any attempt to cheat in discussing such a controversial matter in the most widely read scientific journal would have been instantly apparent. Not only is the failure of Officer and Drake to cite the Snowbird I report a mystery, so is why their oversight was not corrected during the peer-review process.

The Alvarez team accused Officer and Drake of making another scientific error: ignoring explicit warnings from the authors of the original papers on paleomagnetism that their data might be unreliable. Subsequent writers ought not give more credence to data than those who report them in the first place. Officer and Drake began their description of the deep-sea core, which appeared to place the K–T boundary in Chron 29N, by quoting the original paper, where it was described as "the most complete biostratigraphic record of the Cretaceous-Tertiary transition."[7] However, the original authors also said that this particular core had been disturbed during drilling and affected by stirring of muds by burrowing organisms (an effect known as *bioturbation*), both of which would have upset the magnetic patterns, making correct interpretation difficult to impossible.[8] By failing to cite published, negative evidence, Officer and Drake appeared to be unfairly favoring their own position. They also failed to cite detailed magnetic studies from the San Juan Basin that had indicated at first that a particular magnetic reversal was present there that had been found nowhere else, which was so unlikely as to call into question the interpretation of the magnetic reversal data.[9] After further work led to the correction of the discrepancy, the K–T boundary at that site was found to fall in Chron 29R, which Officer and Drake neglected to reveal. This left the Montana magnetic result as the only one remaining from their original set of 15 in which the K–T boundary appeared to reside somewhere other than in Chron 29R. Officer and Drake claimed that "The magnetic stratigraphy observations cover only a short interval of geologic time, but on the basis of faunal correlations with the San Juan Basin, we infer that the reversed interval is 28R."[10] In this case they did refer to the original paper, but failed to note that the original authors had also warned that "The magnetic zones recorded in these terrestrial sections in Alberta, Montana, and New Mexico cannot be securely correlated with the magnetic polarity time scale."[11]

Officer and Drake had offered evidence from three geologic sections that they claimed showed that the K–T event took place at different times around the world and therefore could not have been the result of an instantaneous global catastrophe, which if true would falsify the Alvarez theory (argument 1a). When the original papers that they cited were reviewed in detail, however, their claimed evidence was found either to be nonexistent or to be in serious doubt. (Later work confirmed that Officer and Drake were indeed wrong: Wherever it has been studied, the K–T boundary falls firmly within Chron 29R.)

In a 1984 paper, Walter and Luis Alvarez, Asaro, and Michel had the last word: "A review by scientists who have not been active in the field might have been valuable if it had been balanced, but unfortunately, Officer and Drake use a double standard, in which they apply keen scrutiny to evidence favoring the impact theory—as, of course, they should—but uncritically accept any results, no matter how flawed, that contradict it. They fail to mention most of the data that support the theory. Instead, they fix their attention on a few cases that can be made to look like contradictions."[12]

The title of Officer and Drake's 1983 paper in *Science,* "The Cretaceous–Tertiary Transition," conveyed the message of their argument 1b, showing that the change from the one geologic period to the other had not been sudden, much less instantaneous. In their view, there had been instead a finite "transition," a gradual shift, consistent with uniformitarianism. Under the impact scenario, the time taken by the transition "should be zero," they argued, and if the fossils on one side of the boundary had been gradually replaced by those on the other, rather than disappearing suddenly right at the K–T boundary, the event could not have been instantaneous. But as we will see, the fossil record can be hard to read. Fossil-bearing sediments, after being deposited initially, can be stirred by waves and redeposited, upsetting the original stratigraphy in a process called *reworking*. Burrowing organisms carry material from one stratigraphic level to another, mixing up the sedimentary and fossil record and making once sharp peaks appear gradual (bioturbation). Officer and Drake did acknowledge this difficulty, writing, "Bioturbation is an important process affecting marine sedimentary sequences and can blur or obscure transition events," but never again in their paper did they refer to the process as having any actual effect.[13] Thus they appear to have paid only lip service to bioturbation. Because of the indeterminacy introduced by reworking and bioturbation, pro-impactors regarded the Officer-Drake argument from fossils as completely unproved and no threat to the new theory.

Officer and Drake ended their abstract with this plea: "It seems more likely that an explanation for the changes during the K–T transition will come from continued examination of the great variety of terrestrial events that took place at that time, including extensive volcanism, major regression of the sea from the land, geochemical changes, and paleoclimatic and paleoceanographic changes."[14] The Alvarezes, Asaro, and Michel responded that "Officer and Drake's article seems to be a plea for a return to the time before the iridium anomaly was discovered, when almost any speculation on the K–T extinction was acceptable. This idea is pleasantly nostalgic, but there is by now a large amount of detailed astronomical, geological, paleontological, chemical, and physical information which supports the impact theory. Much interesting work remains to be done in order to understand the evolutionary consequences of the impact on different biologic groups, but the time for unbridled speculation is past."[15]

Thus both counterarguments concerning the timing of the K–T boundary and extinction failed: Officer and Drake were unable to falsify the Alvarez theory preemptively by placing the K–T boundary in different magnetic chrons (argument 1a). Due to the nature of the fossil record, they could not show convincingly that changes in fossil abundances had not taken place in zero time—reworking and bioturbation made it difficult if not impossible to say how long the transition took (argument 1b).

Beyond the disagreements that proponents of the Alvarez theory had with Officer and Drake over scientific facts and interpretations, the failure of the pair to follow standard scientific procedures aroused suspicion of their motives and their modus operandi. In their papers in *Science,* Officer and Drake at times failed to cite relevant papers and contradictory evidence, ignored cautions of primary authors about the reliability of data, used techniques that were more common to debating tournaments than scientific literature, and appeared to pay only lip service to facts that complicated their arguments. The impression created by their methods sowed seeds of distrust that were to bear a bitter fruit.

Officer and Drake soon turned to the second part of their attack: to show that the evidence of impact presented by its supporters—the iridium spike, the shocked minerals, the microtektite-like spherules—were not markers of an extraterrestrial event after all.

IRIDIUM HILLS

Even though later evidence appeared to offer more support for the Alvarez theory than the iridium findings, it was the Gubbio iridium anomaly that sent the Alvarezes down the trail of impact in the first place. If that particular evidence were weakened or falsified the entire theory would be in jeopardy. Iridium had already met two tests: It proved uncommon in the geologic record and, as shown by the presence of the iridium spike in freshwater rocks from the Raton Basin, did not come from seawater. Officer and Drake focused on two other ways the iridium evidence might have resulted from something other than meteorite impact. Their first claim was that iridium was not concentrated in a sharp peak, but spread out above and below the K–T boundary. On a graph of the amount of iridium found at different depths, instead of a sharp peak, there would be a "hill." If such a spread of iridium could be shown not to have been produced by reworking or bioturbation, then the iridium could not have been emplaced by an instantaneous event such as meteorite impact. Second, if it could be shown that normal geologic processes can concentrate iridium, the Alvarez theory would not be required to explain the high iridium concentrations and the way would be open for a uniformitarian alternative.

In their 1983 paper, the two authors had claimed that in some deep sea drill cores that capture the K–T boundary, instead of being concentrated in a spike, iridium is spread over as much as 60 cm (2 ft). They cited the measurements of F. C. Wezel, who had reported high iridium at Gubbio from levels well above and well below the boundary clay.[16] In a black shale 240 m below the boundary, equivalent to millions of years of sedimentation before the K–T boundary, Wezel reported an iridium anomaly twice that of the boundary clay (which, if true, would falsify prediction 2 discussed in Chapter 4). When the Alvarezes, Michel, and Asaro attempted to reproduce Wezel's results, however, they could not. They attributed the discrepancy to contamination in Wezel's laboratory, which they said is "all too easy in chemical analytical work at the parts-per-billion level."[17]

In 1985, Officer and Drake launched a much more broadscale attack on the Alvarez theory. They repeated their earlier claims and upped the ante: "The geologic record of terminal Cretaceous environmental events indicates that iridium and other associated elements were not deposited instantaneously but during a time interval spanning some 10,000 to 100,000 years. The available geologic evidence favors a mantle [volcanic] rather than meteoritic origin for these elements. These results are in accord with the scenario of a series of intense eruptive volcanic events occurring during a relatively short geologic time interval and not with the scenario of a single large asteroid impact event."[18]

Officer and Drake continued to press the claim that the spread of iridium values was too great to be explained by bioturbation, citing evidence of iridium hills rather than spikes from other localities. They again used Wezel's report of high iridium in samples far from the K–T boundary at Gubbio, but they failed to cite the point that the Alvarezes, Michel, and Asaro had made that at least some of Wezel's anomalous iridium levels were due to contamination. Curiously, Officer and Drake did not attempt at all to rebut the charges made in 1984 by the Alvarez team, merely lumping them together with several others under the catchall of "a variety of responses." One coming late to the debate would never have known that the "variety" included many substantive criticisms and an accusation of outright error.

Jan Smit and UCLA's Frank Kyte responded that the Officer and Drake bioturbation model "is inaccurately applied and inadequately explains possible sedimentary effects for any given section."[19] Smit and Kyte describe what once must have been sharp microtektite layers that are now dispersed over an average of nearly 60 cm, showing that bioturbation and reworking can affect far more than a few centimeters and that a stretched-out iridium signature need not falsify the impact theory. Since it was not even certain which mineral phases contained the iridium, it is not hard to think of ways of broadening a once-sharp peak. (1) Before the original sediments that contained the iridium hardened into rock, they might have been stirred by waves and then redeposited, which would have smeared out any originally sharp peaks (reworking). (2) Sediments rich in iridium derived from the impact cloud might have been washed off the continents and into the ocean basins where they would mix with other sediments being deposited there, a process that could have taken hundreds or thousands of years and spread iridium over a vertical distance. (3) Iridium might have been dissolved chemically from its original level in the boundary clay and been reprecipitated up or down the section.

In 1988, my MIT graduate school colleague Jim Crocket, of McMaster University in Hamilton, Ontario, an expert in measuring the concentrations of platinum-group metals, reported a new set of iridium results.[20] With Officer and others, Crocket presented high iridium values that spread for 2 m above and 2 m below the K–T boundary at Gubbio, representing some 300,000 years of sedimentation. The authors ruled out bioturbation by citing their own work, referring to the claim made by Officer and Drake in 1985 that bioturbation affects only 5 cm of rock on the average, far less than the 4-m spread observed.

In the spring of 1988, Robert Rocchia and an international group of pro- and anti-impactors returned to Gubbio to remeasure the magnetic stratigraphy and the iridium distribution.[21] They could not reproduce the high iridium readings above and below the boundary reported by Crocket et al. In 1990, Walter Alvarez, Asaro, and Alessandro Montanari measured a detailed iridium profile across 57 m of the K–T section at Gubbio, which represents about 10 million years of sedimentation.[22] They found an iridium anomaly of 3,000 ppt exactly in the K–T boundary clay, with small molehills of 20 ppt–80 ppt on either side, fading away to the background level of about 12 ppt. Their results essentially matched those of Robert Rocchia and his colleagues.

When responsible authorities come to different conclusions over what is essentially an analytical matter (how much of an element is present in a set of samples), the best procedure is to have the samples analyzed in several independent laboratories, in what is called a blind test, with none of the laboratories knowing the exact derivation of the individual samples. Robert Ginsburg of the University of Miami supervised the collection and distribution of samples from Gubbio. When the results were returned, only a single iridium peak had been found, though it did retain its adjacent shoulders.[23]

This debate reveals the difficulty of saying whether the vertical spread of a chemical signature such as iridium's indicates that the element was deposited with that distribution, as Officer and Drake argued, or whether instead it was deposited in a sharp peak that was later degraded by secondary processes, as the Alvarez team would have argued had they believed the data. To rephrase the question: Is a spread-out iridium hill, as opposed to a sharp peak, a primary or a secondary feature? We certainly know of processes that can degrade a sharp peak into a hill: reworking, bioturbation, erosion and deposition, and chemical solution and reprecipitation. Since 1980 when the Alvarez theory appeared, processes have also been discovered that can remove the iridium naturally present in minute amounts in a section of rock and concentrate it at a particular geologic level, thus turning a broad distribution into a peak. However, these processes do not appear to be able to produce the high iridium levels found at the K–T boundary. Thus it seems safe to conclude that the sharp peaks found at Gubbio and the Raton Basin, for example, are highly likely to be primary and to record an instantaneous and singular event. For this reason, sharp iridium peaks corroborate the Alvarez theory to a far stronger degree than the negative or indeterminate evidence of iridium hills at other locations detracts from the theory. One peak is sufficiently likely to be primary as to be worth several hills.

In 1996, A. D. Anbar and his co-workers at Caltech used ultrasensitive techniques to measure the minute amounts of iridium in rivers and the sea.[24] They found the K–T boundary clay to contain 1,000 times as much iridium as all the world's oceans put together, confirming that the iridium did not precipitate from normal seawater. They also determined that iridium, once present in the oceans, remains there for some 10,000 to 100,000 years before it is removed by sedimentation, providing yet another way to explain the iridium hills: They could merely be the result of the long residence time of iridium once it had been injected into the oceans by meteorite impact.

SHOCKED MINERALS

Bruce Bohor's 1981 discovery of shocked quartz (previously found only at known impact craters and at the sites of nuclear explosions, in the K–T boundary clay in the Hell Creek area of Montana, home of *T. rex*) convinced many geologists that impact was a reality. Glenn Izett of the U.S. Geological Survey, who wrote the definitive paper on the K–T section in the Raton Basin, spoke for them: "I started off as a nonbeliever. What got me was the appearance of these shocked minerals at the K–T. In the impact bed, you see grains everywhere that have these features in them. Just a millimeter or two below, you'll never see any of those features."[25]

Unshocked quartz has no fracture planes; quartz deformed in other geologic settings than impact sites sometimes has single sets. The multiple sets of crisscrossing planes illustrated in Figure 12, however, are diagnostic of great shock. The planes actually are closely set layers of glassy material precisely oriented to the crystal structure of quartz. Officer and Drake took exception to the claim that shocked quartz was diagnostic of impact: "The presence of lamellar quartz features [the parallel planes] does not in and of itself demonstrate a meteor impact origin."[26] As evidence, they stated: "Lamellar features . . . are also a characteristic of both normal tectonic [mountain building] metamorphism and shock metamorphism, although the normal tectonic features are quite different from the

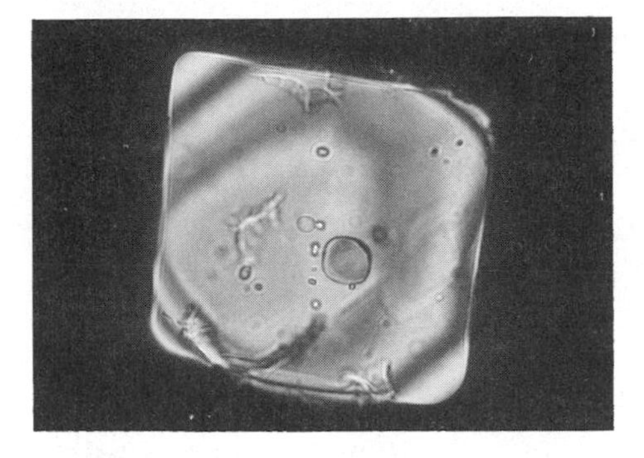

FIGURE 12 (Top) Unshocked quartz from an explosive volcanic rock in the Jemez Mountains, New Mexico. Note the absence of shock planes. (Right) Shocked quartz from the K–T boundary in the Raton Basin, Colorado, showing two sets of shock planes. [Photo courtesy of Glenn Izett.[27]]

shock features," meaning that shocked quartz is produced not only by meteorite impact, but by such familiar geologic processes as metamorphism.[28]

Officer and Drake cite as evidence the presence of shocked minerals at the giant, ancient structures of Sudbury and Vredefort, which they say are of internal, nonimpact origin, and conclude on that basis that shock features do not "demonstrate a meteor impact origin."[29] They state two premises and use them to draw a conclusion: (1) Sudbury and Vredefort were not formed by impact but are of internal origin; (2) both contain shocked minerals; therefore (3) shocked minerals are not diagnostic of impact. If this technique seems familiar, it is. It has a history among debaters and rhetoricians extending all the way back to Aristotle. If the original statement in such a three-step syllogism is itself false, however, then the chain of logic breaks down and the final conclusion may be false. If Sudbury and Vredefort *are* of impact origin, then the conclusion in step 3 could be false (it could also be true but there is no way of knowing from this logic). Based on several lines of evidence, geologists now believe that meteorite impact did create both structures. At least it is sufficiently likely that Sudbury and Vredefort were formed by impact that they cannot be cited as evidence that the recognized impact markers, such as shocked quartz, are not diagnostic of impact.

The Officer-Drake paper drew a strong rebuttal from Bevan French of NASA, an expert on shock metamorphism and the co-editor of a 1968 classic[30] on the subject, who stated: "No shock-metamorphic effects have been observed in undisputed volcanic or tectonic structures."[31] Officer and Drake and their co-workers soon responded, pointing to the giant caldera of the ancient volcano Toba, on the island of Sumatra, near Krakatoa but 50 times larger, which last erupted about 75,000 years ago. In 1986, Neville Carter of Texas A & M, an expert on shocked quartz, along with Officer and others, reported that feldspar and mica in the Toba volcanic ejecta contain microstructures resembling those produced by shock, although the structures were rare in quartz from Toba, and were not the multiple, crisscrossing sets known as planar deformation features.[32]

Bohor and two colleagues from the U.S. Geological Survey then reported shocked quartz at seven additional K–T boundary sites.[33] They also studied quartz grains in the Toba rocks and found that 1 percent show fractures, contrasted with the 25 percent to 40 percent of quartz grains in a typical K–T boundary clay that show the fractures. More importantly, the deformed quartz from Toba exhibits only single sets of parallel planes, not the multiple planar deformation features characteristic of impact shock.

A group of Canadian geologists compared quartz grains from Toba, a known impact site, the K–T boundary clay, and two sites known to have undergone tectonic deformation. They found that the appearance and orientation of planar features from the known impact structure and those observed in samples from the K–T boundary were essentially identical. They concluded that although other lamellar deformational features in quartz can result from other geologic processes, these features only superficially resemble those from the K–T boundary and those believed to have resulted from impact.[34]

According to an article by Richard Kerr, a reporter for *Science* magazine who has covered the meteorite impact debate from its inception, Neville Carter agreed, saying that "there is no question that there is a difference."[35] Kerr noted that Carter found "no quartz lamellae whatever in distant Toba ash falls."[36] In his published rebuttal with Officer, however, Carter appeared to reverse himself: There, their evidence was said to "clearly repudiate *all* [italics theirs] assertions of Alexopoulos et al."[37]

Such claims did not persuade Kerr, who summed up: "Try as they might, advocates of a volcanic end to the Cretaceous have failed to find the same kind of so-called shocked quartz grains in any volcanic rock. Because shocked quartz continues to maintain its exclusive link to impacts, the impact hypothesis would seem to be opening its lead over the sputtering volcanic alternative."[38]

In 1991, Officer and Carter published a lengthy review paper on what they called "enigmatic terrestrial structures."[39] Although they did admit that some of the cryptoexplosion structures studied by Walter Bucher and others were due to meteorite impact, they concluded that the Sudbury and Vredefort structures, and several others whose origin had been disputed, were of "relatively deep-seated," that is, internal, origin. One of the features presented as of internal origin was the buried structure at Manson, Iowa, which as we will see became a prime contender for the K–T impact site and is now firmly regarded as caused by impact. After an extensive review of shock metamorphism, Officer and Carter wound up their argument:

> Perhaps the widely held, but erroneous, belief that only multiple sets of these features are diagnostic of shock deformation has resulted from their many recent illustrations at the Cretaceous/Tertiary boundary. However, surprisingly, this myth has also been promulgated recently by workers with extensive experience with dynamic deformation of quartz. The experimental work summarized above should be sufficient to convince the critical, unbiased reader that single sets of planar features are just as diagnostic of shock deformation as multiple sets. This information combined with the observation that single sets are just as common as are multiple sets in naturally shocked quartz should finally put this nontrivial matter to rest.[40]

It appears to be true that single sets of deformation planes indicate some level of shock. As the debate went on, single sets were found in other volcanic rocks, including ash from Mount St. Helens. But it is also true that the *multiple*, crisscrossing sets of deformation lamellae—the planar deformation features—found repeatedly in quartz at the K–T boundary, occur only at known impact structures, in high-pressure laboratory experiments, and at the sites of nuclear explosions. Volcanic rocks do not contain them, and therefore the multiple sets of planes remain diagnostic indicators of a higher level of shock than produced by volcanism—or, as far as we yet know, by any internal process. To quote three experts: Impact-shocked quartz and quartz altered by other terrestrial processes "are completely dissimilar . . . due to the vastly different physical conditions and time scales. . . . Well characterized and documented shock effects in quartz are unequivocal indicators of impact."[41]

MYSTERIOUS SPHERULES

While working on the iridium-rich K–T boundary clay at Caravaca, Spain, in 1981, Jan Smit and G. Klaver discovered rounded, sand-sized grains of feldspar.[42] Similar spherules showed up at the other prominent K–T sites and in several deep-sea cores that captured the boundary. On the basis of the mineralogy and texture of the spherules, Smit, Alessandro Montanari, the Alvarezes, and their colleagues concluded that they were congealed droplets of molten material that had been blasted aloft in the K–T impact explosion.[43]

Wezel and his group reported that, like iridium, the spherules spread out above and below the K–T boundary at Gubbio, undercutting the claim for a special event at the boundary.[44] Officer and two colleagues from Dartmouth wrote that they had found spherules over a vertical span of rock at Gubbio equivalent to 22 million years, analogous to their alleged findings of spread-out iridium.[45] The Wezel claim led to one of the more bizarre episodes in the debates over the Alvarez theory, which is saying a lot. Officer presented Wezel's results at the 1985 meeting of the American Geophysical Union, but Montanari, who had also collected and studied the Gubbio spherules, came up with quite a different interpretation of what Wezel and company had found, which he then shared with his colleague, Walter Alvarez. Let us pick up the story in the words of the protagonists, as reported by Malcolm Browne in the *New York Times* of January 19, 1988:

> But according to Dr. Alvarez, "My son Walt took just two minutes to demolish Officer after he delivered that paper." Dr. Alvarez said his son showed that the "spherules" found by Dr. Officer's team were merely insect eggs and had been mistaken for mineral spherules because they were not cleaned well enough. "At that point," Dr. Alvarez wrote in his autobiography, "the audience of several hundred Earth scientists burst into laughter, something I'd never witnessed before in my 53 years of attending scientific meetings."
>
> Dr. Officer responded: "This is a misstatement. There was no outburst of laughter following Walter's brief comment, and no direct or implied derision of me as a scientist by the audience." "My talk at that meeting," he said, "concerned the hypothesis that intense volcanic activity and the lowering of sea levels explains the mass extinctions at the end of the Cretaceous. During that talk, mention was made of the distribution of microspherules. Walter had kindly pointed out to us previously that there were contaminant hollow spherules of recent origin as well as solid spherules of a mineral composition indigenous to the geologic section."

"After duly eliminating the insect eggs and giving due credit to Walter in our subsequent scientific publications," Dr. Officer said, "we found that all the solid spherules, throughout the whole section, extended above and below the terminal Cretaceous layer. They were present in sediments spanning a time period of several million years and could therefore not have come from impact."[46]

As research continued, the spherules turned up at over 60 K–T sites. Those found outside the United States were solid, rich in nickel and iron, and often composed of spinel. Some of the spinels have iridium concentrations of up to 500 ppt. These spinels are not a dispersed chemical element like iridium, which for all we know can be dissolved and reprecipitated, and whose exact source in the boundary clay is not known to this day, but rather are physical objects—spherules up to 100 microns (1 micron = 10^{-6} m) in diameter. Once locked into a sediment, these spherules would be difficult to move and thus they help to decide whether the strange features of the boundary clay—iridium, shocked quartz, spherules—were originally present in a peak or in a hill. A group of French geologists found that at El Kef, Morocco, and at several other K–T sites, iridium spreads over a broad hill, but the spinels occur in a razor-sharp peak right at the boundary.[47] They concluded that the spinels (and thus the boundary clay) took less than 100 years to deposit—a blink of the eye in geologic time.

CHALLENGE MET

Officer and Drake succeeded neither in their effort to falsify impact by showing that the K–T event was not instantaneous (arguments 1a and 1b) nor in their attempt to discredit the evidence for impact (argument 2). In the process, though, scientists learned a great deal, especially about the geochemistry of iridium. Certainly the efforts of the doubters failed to discourage the proponents, who were growing in number. But on the other hand, those who supported the theory were equally unable to sway its firmest opponents. In fact, only a vanishingly small number are on record as *ever* having changed their minds on the Alvarez theory. One need read only a fraction of the vast literature on impact to predict with near certainty which side a given author will take in all subsequent papers: the same as in previous ones. Glen notes that he has "found neither in planetary geology nor in impacting studies anyone who ever wavered from the impact-as-extinction-cause component of the hypothesis, nor in vertebrate paleontology anyone who converted to embrace it."[48] Michael Rampino of New York University and NASA; paleontologists Leo Hickey and Kirk Johnson of the Denver Museum of Natural History; and Peter Ward of the University of Washington, plus a handful of others, are exceptions, but they merely prove the rule. Some of the reluctance to switch sides is undoubtedly due to honest convictions firmly held, but some also results from the unwillingness of scientists, being human, to admit in public that they were wrong. And the role of tenacious skeptic, adhering faithfully to the old ways that have served so well for so long, can be a proud one. Even if eventually proved wrong, one fought the good fight and can hold one's head high. The trick is not to fight too long, or unfairly.

The critics of the impact theory next turned to argument 3: to replace the Alvarez theory by showing that another process—one as familiar to geologists as an old shoe—explains the evidence equally well and obviates the need for a deus ex machina. If they could not convince the pro-impactors of the error of their ways, at least the anti-impactors could present a persuasive case that would shore up support among those who had not yet made up their minds. The Alvarez theory would then eventually be discarded in that large dustbin of discredited theories.

Chapter 6
The Volcanic Rival

Is he in heaven? Is he in hell?
That demmed, elusive Pimpernel?[1]
Baroness Orczy

In 1972, Peter Vogt, a volcano specialist at the U.S. Naval Research Laboratory in Washington, D.C., called attention to the huge volcanic outpourings that had occurred in India at the time of the K–T boundary and wondered if the resulting injection of poisonous trace elements into the atmosphere might not have been the cause of the mass extinction.[2] Volcanism indeed makes an attractive rival to the Alvarez theory, as it is the only process other than impact that meets the dual criteria of being lethal and global. Furthermore, volcanoes erupt today and it is easy to project their effects backward in time. In conformity with Hutton's teachings, present volcanic activity might well be the key to past extinction.

In 1978, Dewey McLean of Virginia Tech proposed that the carbon dioxide accumulations at the end of the Cretaceous had caused changes in oceanic circulation and global climate (perhaps a kind of greenhouse effect) that in turn led to the mass extinction.[3] In 1985, Officer and Drake adopted and refined these arguments, claiming that the iridium, shocked minerals, and spherules found at the K–T boundary are more likely to have been formed by volcanism than by impact.[4] Although a single volcanic eruption can be almost as sudden as impact—witness the explosion of Mount St. Helens in 1980—it takes hundreds of such eruptions spread over hundreds of thousands or millions of years to build up a volcanic cone. If Officer and Drake are correct that the supposed impact markers can be produced by volcanism, the alleged spread of the markers for several meters above and below the K–T boundary would then be naturally explained. Thus, to the Officer-Drake school, volcanism tied together all the facts remarkably well.

There was no denying that they were onto something. After the eruptions of Krakatoa in 1883, Mount St. Helens in 1980, and Pinatubo in 1991, the dust and sulfur injected into the atmosphere shaded the earth enough to cause average world temperature to drop, though by only one or two degrees. And remember that it was the report of the darkening effects of the Krakatoa eruption that set Luis Alvarez on the trail of meteorite impact in the first place. It would be the height of irony were Krakatoa now to be used to defeat his theory.

VOLCANIC IRIDIUM

The picture of iridium anomalies as uniquely diagnostic of meteorite impact began to cloud in the mid-1980s, lending additional credence to the volcanism theory. The chemistry of aerosols (suspensions of fine solid or liquid particles in gases) emitted from Kilauea Volcano in Hawaii had been under investigation by scientists from the University of Maryland.[5] Although for five years they detected no iridium, aerosols from the 1983 eruption unexpectedly contained up to 10,000 times as much iridium as the Hawaiian basalts. Officer and Drake pointed out that the iridium in the airborne particles was "comparable to concentrations associated with meteorites."[6]

The picture quickly clouded further: High iridium levels were discovered in particles emitted by a volcano on the remote island of Réunion in the western Indian Ocean[7] and in the ejecta of silicic volcanoes on Kamchatka.[8] Iridium levels as high as 7,500 ppt, comparable to the K–T levels, were found in layers of volcanic dust buried in the Antarctic ice sheet.[9] Thus, contrary to the view that prevailed when the Alvarez theory was first introduced, as the 1980s progressed it began to appear that certain volcanic processes can concentrate iridium and in amounts approaching K–T boundary levels.

In 1996, Frank Asaro, an original member of the Alvarez team, and Birger Schmitz of the University of Gothenburg in Sweden, reported iridium measurements in a number of ash deposits, including some near the K–T boundary.[10] They confirmed the discovery that some types of explosive volcanism produce ash with up to 7,500 ppt iridium, but said that by comparing the levels and ratios of various chemical elements in volcanic rocks and meteorites, it is easy to tell that these were terrestrial iridium anomalies rather than impact-related ones. They found no iridium in the types of ashes studied by the Russian geologists, whose claim, they therefore said, needed further confirmation. They thought that the Antarctic iridium, rather than stemming from volcanism, might be derived from the meteoritic dust that has settled there for millennia. The other clay layers they studied contained no iridium.

For the volcanic alternative to be viable, volcanoes must have emitted large enough volumes of lava to allow their by-products, such as carbon dioxide or dust, to cause a mass extinction. Those by-products would have to include the iridium, shocked quartz, and the spherules found at the K–T boundary, all of which the volcanoes would have to distribute around the globe. The difficulty is that although all types of volcanoes taken collectively might explain these observational facts, none of the individual types do. We know that volcanoes such as Krakatoa, and those of the Ring of Fire—the group of active volcanoes that encircle the Pacific Ocean basin from Tierra del Fuego around to the Philippines—explode suddenly and unexpectedly. They do so because the chambers beneath them hold a volatile mixture: magmas (subterranean lavas) rich both in silica and in gases kept in solution under high pressure. These silicic magmas are thick and viscous, like molasses, which causes them to clog their volcanic conduits, trapping the dissolved gases. The gases can then burst free in a gigantic explosion, like a too-rapidly opened bottle of carbonated beverage, shooting plumes of dust and ash into the stratosphere and showering debris for thousands of kilometers. In 1980, Mount St. Helens blasted itself to pieces in an explosion that sent fine ash wafting over most of the United States. By the time it reached the eastern states, however, the heavier fraction of the ash had already settled out, leaving suspended a portion so fine as to be almost invisible.

K–T quartz grains and spherules are much larger and heavier than fine volcanic ash. Had they erupted into the stratosphere, they would quickly have fallen back to earth. No one has been able to show how explosive volcanism can send large particles winging around the globe; in any case, volcanic explosions produce angular glass shards, not rounded spherules. And as noted in Chapter 5, the multiple, crisscrossing planar deformation features common in boundary clay quartz have never been found in volcanic products (including the quartz from Mount St. Helens).

The high iridium levels measured in volcanic aerosols from Hawaii came from a different type of volcano than those of the Ring of Fire. The Hawaiian variety erupts basalt, which is lower both in silica and in dissolved gases than the lavas of the Ring of Fire, making it less viscous and more able to flow. For this reason, basaltic eruptions are quiescent rather than explosive and their lavas are restricted to the nearby area. Although gases emitted from these basaltic volcanoes might convey iridium around the globe, the lavas themselves contain almost none. Because basaltic lavas erupt quietly, it is hard to see by what process they could produce the required worldwide distributions of iridium, shocked minerals, and spherules, especially since basalt contains negligible iridium and no quartz. Thus basaltic volcanism also fails to explain all the evidence.

Although silicic volcanoes contain quartz and explode, they emit smaller volumes of material, and for more limited periods of time, than the basaltic variety. The famous eruptions of Toba, Krakatoa, and Mount St. Helens did not come close to causing a mass extinction. The only volcanoes known to erupt large enough volumes of lava over a long enough period of time to produce potentially lethal amounts of chemicals and cause a global mass extinction are basaltic, yet basaltic volcanoes emit their products so quietly that they do not receive worldwide distribution. It is hard to put all this together into a satisfactory substitute for meteorite impact. Nevertheless, one of the most massive outpourings of basalt in earth history did erupt in India close to the time of the K–T boundary, a worrisome coincidence for the pro-impactors. Another is that the greatest mass extinction of them all—the one between the Permian and Triassic periods—occurred at nearly if not exactly the same time as a huge outpouring of basaltic lava in Siberia.

FLOOD BASALTS

The Indian Deccan traps occur over an area of at least 1 million km^2. (*Deccan* is Sanskrit for southern; *traps* is Swedish for staircase, which the edge of a giant sequence of nearly horizontal lava flows sometimes resembles.) In places they are 2 km thick. Their total volume exceeds 1 million cm^3, more than the outpourings of all the Ring of Fire volcanoes put together. Scientists believe that the Deccan eruptions produced 30 trillion tons of carbon dioxide, 6 trillion tons of sulfur, and 60 billion tons of halogens, gases that enhance the greenhouse effect.

To reflect their vastness and mode of eruption, geologists call these enormous outpourings flood basalts.[11] They occur in the geologic record from the Precambrian to the Tertiary and on nearly every continent. Examples are the Columbia River country of Oregon and Washington, the Paraná basin of Brazil, the South African Karroo, and the Siberian traps. Before the advent of plate tectonics, their origin was a mystery, but Tuzo Wilson, tolerated but unheeded in my graduate school seminar, provided the key insight.

Wilson was a master at providing an innovative interpretation of facts that had stumped others, in this case, that the Hawaiian Islands lie along a straight line. Wilson noted, as had many, that the islands to the southeast, where Kilauea and the other great volcanoes are active today, are youngest and that they grow steadily older in a line to the northwest. Beyond the most northwesterly island lies a linear chain of submarine seamounts that become progressively deeper and older, also to the northwest. Unlike the many others who knew these facts, Wilson deduced their meaning. Deep in the earth's mantle, beneath the present active volcanoes, lies a *hot spot*, a zone that melts periodically, sending jets of less-dense magma up to be extruded onto the surface. But plate tectonics tells us that the rigid, uppermost surface layer of the earth moves horizontally over the fluid mantle, so that as time passes different sections of the crust lie over a given spot in the mantle deep below. Now we can begin to see what Wilson envisioned: A fixed hot spot deep beneath the central Pacific crust episodically spurts magma toward the surface, but by the time each spurt arrives, the crust has moved laterally so that the new eruption occurs at a different point on the surface, producing a new volcano. The former volcanic sites move progressively further and further away from the point directly above the hot spot, becoming older, eroding, and finally disappearing beneath the sea. From this, Wilson deduced that the crust in the mid-Pacific has been moving steadily to the northwest, over the fixed position of the deep hot spot. Since the ages of the Hawaiian volcanoes and some of the seamounts are known precisely, we can calculate the speed at which the crust there is moving; the rate checks exactly with results obtained using the magnetic reversal time scale.

Jason Morgan of Princeton extended Wilson's idea to explain the origin of flood basalts. When he reconstructed the past positions of tectonic plates, he noticed that the basalt provinces lie directly over present-day active volcanoes. For example, the hot spot that is now under Yellowstone National Park once produced the Columbia River basalts. Morgan thought that hot spots at the base of the earth's mantle sometimes produce huge, bulbous masses of hot, low-density, low-viscosity basaltic magma, that then "float" to the top of the mantle, like a hot air balloon rising through colder, denser air. There these giant mushrooms of magma rapidly decompress and flow out on the surface. Some think all this happens in only a few million years. Thus hot-spot eruptions can produce not only midoceanic islands but, perhaps, giant floods of basalt.

In the late 1980s, the Ocean Drilling Program explored the western Indian Ocean and found a chain of seamounts extending from southwest India to the island of Réunion. The ages of the seamounts ranged from 2 million years near Réunion to around 55 million to 60 million years just south of India. A reconstruction of movement of the plates in the Indian Ocean area shows that at the end of the Cretaceous, the Deccan trap province resided just over the hot spot that is now feeding Réunion.

THE DECCAN TRAPS

The sharp iridium peaks, the spinel spikes described in the last chapter, and the thin boundary clay tell us that whatever the origin of the K–T event, it did not last for millions of years, or even for 1 million years. As we have seen, some think it lasted for no more than 100 years! If Deccan volcanism caused the K–T event, the eruption of the traps must have started just prior to K–T time and lasted just beyond it, covering at most a few hundred thousand years. If the Deccan traps do not date exactly to the K–T boundary, or if they erupted over several millions of years, the volcanic alternative would itself be falsified.

Thus the crucial question is this: When did Deccan volcanism begin and how long did it last? One way to answer the question is to use the magnetic reversal time scale. Frenchman Vincent Courtillot and colleagues measured the magnetic reversal age patterns of the Deccan basalts and concluded that the eruptions began during Chron 30N, reached a maximum in Chron 29R, and were waning by the time of Chron 29N.[12] This means that Deccan volcanism could have lasted for 1 million to 2 million years (see magnetic reversal time scale in Figure 11). Chron 29R is the time period during which the K–T event, whatever caused it, took place.

As noted before, magnetic reversal dating suffers from an inherent lack of precision, generally coming no closer than a few hundred thousand years. The ages that had been reported in the older literature for the Deccan traps were also anything but precise, ranging from 80 million down to 30 million years, a spread far too great to be plausible and now believed to have stemmed from the difficulty of applying the potassium-argon dating method to altered basalts. A newer technique, the argon-argon method, is more precise and has largely supplanted the older method. Its use narrowed the wide range of the older age results, leaving a spread that was small but real, from 67 million years down to 62.5 million years. As more measurements have been made, this spread has held firm.

To determine whether the Deccan eruptions have the same age as the K–T boundary, we must know the exact age of that boundary. So far, I have simply stipulated that the K–T event took place 65 million years ago, but without presenting any evidence. How is it that we date geologic boundaries precisely? As noted earlier, we cannot do so by using fossils, for they provide only relative ages. We cannot use the magnetic reversal scale because not only does it give a range of ages, ultimately it must tie back to radiometric dating using pairs of parent and daughter atoms, the only method to give absolute rather than relative ages. In short, to date the K–T boundary precisely we must find rocks and minerals from that time whose ages can be measured by one (for comparison ideally several) of the radiometric techniques.

The great interest in the Alvarez theory naturally placed the then-existing estimates of the age of the K–T boundary under close scrutiny and led to a new set of measurements. Beginning in the mid-1980s, several precise analyses were made using different parent-daughter pairs; the results clustered closely around 65.0 million years. Thus we can say with rare assurance that the age of the K–T boundary is 65.0 million years, plus or minus a few hundred thousand. The K–T is surely the best dated of any of the major boundaries of the geologic time scale.

INDIAN IRIDIUM

By the mid-1980s, even the fiercest opponent of the Alvarez theory had to admit that the iridium anomalies, however much they spread above and below the K–T boundary and whatever their cause, did mark the position of the K–T boundary and must reflect a global event. If Deccan volcanism was the source of that iridium, it might then be possible to find an iridium-rich K–T boundary layer amidst the Deccan basalt flows. An intensive search for iridium by the French team came up empty, however. Finally a group of Indian geologists, led by N. Bhandari of the Physical Research Laboratory in Ahmedabad, discovered the iridium needle in the Deccan haystack.[13]

The basalt flows that make up the traps alternate with layers of sediment called *intertrappeans*. The alternating stratigraphy of the basalts and the intertrappeans tells us that volcanism began, stopped long enough for sediments to accumulate, resumed, stopped again while more sediments collected, and so on until a layer cake of basalt and sediment built up.

The Indian geologists sampled in the Anjar region of Gujurat State, where seven basalt flows are recognized, each separated from the next by intertrappean sedimentary layers several meters thick. The third intertrappean bed from the bottom, ITIII, contains bones and eggshells of dinosaurs. The Indian geologists used the argon-argon method to date the lava beds designated FIII and FIV, which lie above and below ITIII, at 65.5 ± 0.7 and 65.4 ± 0.7 million years, identical to the 65.0-million-year date for the K–T boundary. (This evidence also shows that dinosaurs were still alive up to the very end of the Cretaceous.) Since the third lava bed is 65 million years old, the two earlier ones must be older, as the magnetic reversal results indicated they are, confirming that Deccan volcanism started well before the K–T boundary.

Within sedimentary layer ITIII, just above the highest dinosaur fossil, there are three chocolate-brown layers each less than a centimeter thick. In one of these thin layers, Bhandari and colleagues found a sharp iridium peak reaching 1,271 ppt, compared to a background of less than 10 ppt in the basalts (the low levels in the basalt were later confirmed by Schmitz and Asaro) and to less than 100 ppt in the nearby intertrappean sediments. Osmium levels are high in the layer and the osmium:iridium ratio is the same as in meteorites and the mantle. The remarkable perseverance and skillful detective work of the Indian geologists clearly confirm that this thin, unremarkable layer is the K–T boundary clay. Even when you know where to look and have good age control, finding an iridium-rich ejecta layer is difficult.

The exemplary findings of the Indian geologists and the accumulated knowledge of the chronology of the Deccan traps lead to several conclusions:

- The magnetic results obtained by the French geologists show that the Deccan eruptions began at least 1 million years before K–T time and lasted for at least 1 million years after it, far too long an interval to be consistent with the considerable evidence that the K–T event was rapid.

- Contrary to the claims of some paleontologists, and others who have opposed the Alvarez theory, the dinosaurs did not die out well before the K–T boundary but lived right up to it. (And some claim they lived on into the Tertiary!)

- Since three Deccan trap flows lie below the layer that contains dinosaur remains, the dinosaurs, and presumably the other species that were exterminated in the K–T event, survived at least the first few phases of Deccan volcanism. Thus the eruption of the Deccan volcanoes was not immediately inimical to life, even when the volcanoes were right next door.

- Some geologists have speculated that a major meteorite impact might have released so much energy that the earth's mantle below ground zero melted, initiating a period of volcanism. In this view, impact might have precipitated Deccan volcanism. But that idea does not work for the Deccan eruptions, which began at least a million years too early.

- Since the K–T boundary is located near the middle of sedimentary layer ITIII, which itself was deposited well after the volcanism that produced basalt flow III had ceased, it is hard to understand how Deccan volcanism could have been the source of the K–T iridium—the iridium was deposited after the Deccan volcanoes had stopped erupting (though they did resume).

- Since none of the other intertrappeans have high iridium, the element apparently was not produced in the normal course of Deccan volcanism and intertrappean sedimentation. Its presence in one thin layer among many suggests, if it does not demand, that the iridium has a special origin, unconnected with Deccan basalts and sediments. The extremely low concentrations of iridium in the Deccan basalts—among the lowest levels ever measured—make them a most unlikely source.

CAREER DAMAGE

By the mid-1980s, not only did the volcanic theory appear to be losing out to the Alvarez theory, some of its proponents felt that they had been treated unfairly by the media. In 1993, Dewey McLean, who has done as much as anyone to develop the volcanic alternative, sent *Science* a complaint that the magazine had shown "indefensible favoritism toward the asteroid and virtual censorship of the volcano extinction theory. Since 1980, *Science* has published 45 proimpact manuscripts and Research News articles and four strictly nonimpact items."[14] Dan Koshland, editor of *Science*, responded that "For our peer-reviewed papers . . . 'freedom of speech' cannot mean 'equal space' for all points of view."[15]

Walter Alvarez passes over these and other unpleasant aspects of K–T debates, saying that "The field as a whole did reasonably well in maintaining a civilized level of discourse."[16] Anti-impactors such as Officer and McLean, and many others, would surely disagree. Walter's claim contrasts vividly with McLean's poignant open letter to Luis Alvarez, and with other information provided on McLean's web page.[17]

According to McLean, Luis Alvarez tried to destroy McLean's career starting at one of the first K–T conferences:

> Luis Alvarez's response was to take me aside at the first coffee break and threaten my career if I opposed him publicly. I had written the first paper showing that greenhouse warming can trigger global extinctions (for the K–T). Alvarez warned me on what happened to a physicist who had opposed him: "The scientific community pays no more attention to him." Alvarez followed through on his threat.
>
> That situation devastated me. By my own originality, I was a principal in a great scientific debate with one of the world's most creative living geniuses, himself working in an environment predicated upon creativity, and I had been undermined, and nearly destroyed, in my own! The stresses over the damage to my career here at VPI did its work. Throughout 1984, nearly every joint in my body was so inflamed, and swollen, that any movement was excruciating; medication kept me nauseated.
>
> Vicious politics by Alvarez, and some paleobiologists, were injected into my department, and used to undermine me in the early–mid 1980s. They nearly destroyed my career, and my health. I developed a Pavlovian-type response to the K–T such that from the mid 1980s until the 1990s, I had great difficulty doing K–T research. My health was so damaged that I was never able to recover, and had to retire in May, 1995.[18]

In his 1988 interview with Malcolm Browne of the *New York Times,* Luis Alvarez said, "If the president of the college had asked me what I thought of Dewey McLean, I'd say he's a weak sister. I thought he'd been knocked out of a ball game and had disappeared, because nobody invites him to conferences anymore."[19]

VOLCANISM DENIED

In Chapter 4 we saw how the meteorite impact theory met several important predictions, and in Chapter 5 how it avoided falsification by the Officer-Drake school. In this chapter, we have seen that the evidence fits an impact scenario better than a volcanic one. If the K–T crater could be found, then, as was said about the meteorite-impact theory in its early days, the volcanic alternative would not be "required." Finding the crater would establish impact as an observational fact and ice the cake of the pro-impactors. As G. K. Gilbert put it long ago when he set out to confirm that impact had created Arizona's Meteor Crater, it is time to "hunt a star."

Chapter 7

To Catch a Crater

It is of the highest importance in the art of detection to be able to recognize out of a number of facts which are incidental and which are vital.[1]
Sir Arthur Conan Doyle

The Alvarezes and their supporters had recognized from the start that finding the K–T impact crater would clinch the impact half of their theory, but they also knew that the odds against finding it were high. For one thing, an incoming meteorite has about a 67 percent chance of striking in the ocean, where, if the resulting crater did not lie in the 20 percent of the seafloor that has disappeared down the deep-sea trenches since K–T time, it is likely to have been covered by younger sediments. If the meteorite struck near one of the poles, the crater might now be covered with ice. In short, it was far easier to think of reasons why the K–T crater, if it existed, should not be found than reasons why it should. Clever detective work, and even more importantly, good luck, would surely be required.

Clues

Fortunately, the crater detectives had the benefit of some important clues. At first, the clues seemed to point toward an oceanic location for the crater, a worrisome possibility given the likelihood that in that case it would no longer be visible. As the K–T boundary clay continued to be studied, however, other evidence began to come to light that suggested a continental landing site. For example, the boundary clay contains a small but persistent fraction of broken rock fragments, the majority of which are granitic, and granite is only

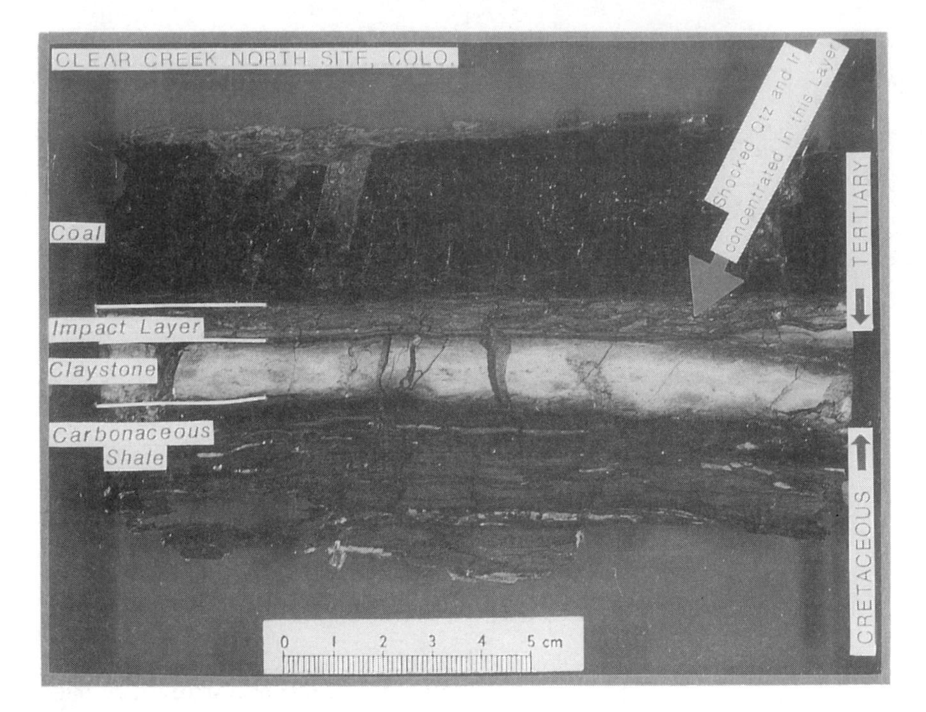

FIGURE 13 The K–T boundary in the Raton Basin. [Photo courtesy of Glenn Izett.]

found on continents. The chemistry of the Haitian glasses, as analyzed by Glenn Izett and his colleagues, and the ubiquitous grains of shocked quartz in the boundary clay, also pointed landward. Quartz is an essential mineral in granites but is absent in the basalts of the ocean floors. By the mid-1980s, quartz grains in the boundary clay at North American sites were recognized to be both larger and more abundant than those found elsewhere, suggesting that the crater was to be found on or near North America. Based on the evidence from the first reports of shocked quartz in rocks from Montana, in 1984 impact specialist Bevan French made the bold prediction that the target must lie no more than 3,500 km from the Montana site.[2]

The North American K–T boundary sites proved to be different in another way, for at them not one but two boundary layers are often found, as shown in Figure 13. The upper layer, about 3 mm thick, contains the iridium spike, shocked quartz, and spherules; the lower layer, about 2 cm thick, carries much less of all three. This double layering, first recognized by Jan Smit, turns out to occur in many of the K–T sections from North America, but nowhere else.

Smit, Walter Alvarez, and others attribute the two layers to two trajectories of impact explosion. Immediately after impact, the explosive fireball, composed mostly of superheated vapor and mineral dust, was lofted high above the stratosphere. Like the volcanic dust from Krakatoa, it took several months to settle back to the earth; arriving last, it became the upper layer. The coarser material, ejected next from the crater, traveled on ballistic trajectories that did not carry it to high altitudes. Thus it settled out first and became the lower layer. The presence of the double layer only in North America was yet another indication that the crater is located there.

The upper stratum shown in Figure 13 is the K–T clay layer on which we have focused so far—the one marking the boundary at Gubbio and the other sites outside North America. It occurs sandwiched within a variety of marine and nonmarine sediments that were deposited at greatly differing rates. Yet regardless of how rapid the rate of sedimentation of the rocks on either side, the clay layer always has the same thickness: 2 mm to 4 mm. This can only mean that it was deposited independent of normal sedimentation processes and at a much faster rate than that of any of the surrounding sediments.[3] Comparison with the settling rates of atmospheric dust suggests that deposition of the upper layer took place over only two to three months.

Manson

In proposing that the target was within 3,500 km of Montana, Bevan French went on to identify two possible craters, one of which was the Manson structure in Iowa. Because it is covered by 30 m to 90 m of glacial debris, Manson is not visible at the surface. Such hidden geological structures are detected using geophysical techniques that rely on magnetism, gravity, and seismic waves. For example, rocks with more magnetic minerals than average produce a positive magnetic anomaly; those that are more dense give rise to positive gravity anomalies. Using such methods, geophysicists can tell a great deal about rock structures that they cannot see; indeed, this is the way they discover buried structures that may contain oil. The geophysicists confirmed that the Manson structure is about 35 km in diameter, smaller than the predicted size of a crater resulting from the impact of a 10 km projectile. On the other hand, Manson is of late Cretaceous age and the basement rocks in the area are granitic. Although Officer and Drake concluded from their analysis that impact did not create the Manson crater (as they also but incorrectly concluded for Sudbury and Vredefort), the discovery of shocked quartz with planar deformation features at Manson, as well as the overall form of the structure, showed that it should be added to the lengthening list of terrestrial craters. Recent seismic studies show that it has a structural central peak (not visible at the surface) nearly 3 km high.[4]

As the attention of the crater hunters turned increasingly to North America, Manson loomed as the natural candidate. Their interest seemed to be justified when the first argon-argon age determination from Manson returned an age of 65.7 ± 1.0 million years, a range that included the age of the K–T boundary.[5] As Manson drew more attention, however, its small size continued to cast doubt. The 10-km impactor predicted by the Alvarezes on the basis of the worldwide iridium levels would create a crater five times Manson's size; one no larger was unlikely to have been able to produce the observed impact effects. Doubt increased when the Izett group obtained additional drill core samples from Manson and dated (by the argon-argon method) unaltered feldspar grains that appeared to have crystallized from the impact melt and which therefore should give the structure's true age. The age came back not at 65 million years, but at 73.8 ± 0.3.[6] The discovery that Manson is normally magnetized and thus cannot belong to K–T Chron 29R confirmed that it is not of K–T age. To pin the matter down, Izett and colleagues, in a neat piece of work, journeyed to nearby South Dakota, where, at the stratigraphic level equivalent to an age of 73.8 million years, they found a zone of shocked minerals.[7] Thus as more evidence has accumulated, Manson has been confirmed as an impact crater, but one that formed at a time different from the K–T.

How to Recognize an Impact Crater

If the K–T crater was located on a continent and easy to spot, it would have been discovered long ago; if it exists, it must either be so eroded or covered by younger sediments, as at Manson, that it is detectable only through the use of geophysical methods. Unfortunately, geologists could not turn to earthly examples to learn how to recognize such huge and obscure structures, for it is rare to find terrestrial craters larger than 100 km. Those that do occur, as at Sudbury and Vredefort, are ancient, distorted, and eroded. Other bodies in the solar system, however, provide ready examples. As exemplified by Tycho (see Figure 7), the moon has many complex craters, with central peaks, collapsed rim terraces, and internal zones of melt rock and impact breccia. Their ejecta can be seen to be scattered over hundreds or thousands of kilometers. From studying smaller craters on the earth, and larger ones on other bodies in the solar system, geologists have built up an accurate and detailed picture of what the K–T crater and its associated features would be like.

- If the impactor were 10 km in diameter, as the Alvarezes calculated, then based on studies of nuclear explosions and cratering on other bodies, the crater would have a diameter of about 150 km. However, if some of the impact material had been blasted completely out of the earth's gravity field, then the boundary clay would not contain all the ejecta, and the resulting calculation would give too small a size for the meteorite and the resulting crater.

- Magnetic, gravity, and seismic anomalies will reveal a circular, bull's-eye pattern and a buried central peak.

- If the crater is not too deeply buried, it might have an obscure but recognizable surface topography, perhaps represented by concentric, arc-shaped ridges and valleys.

- The structure will contain impact breccia and once-molten rock. (However, volcanic rocks were also once molten and can be mistaken for impact melts.)

- The melt rock will be enriched in iridium, display reversed magnetism, and have a radiometric age indistinguishable from 65 million years.

- Glass spherules resembling tektites and dating exactly to the K–T boundary will be found in the vicinity.

- Ejecta layers located farther away will contain shocked minerals and iridium. If they can be dated, they too will be 65 million years old.

- If the impact occurred in the oceans near a continent, say on a continental shelf, it might have given rise to giant waves that would leave unusual sedimentary rocks behind.

This is a long list. Were any putative K–T impact structure to meet even most of these predictions, the search would be over. A site that met them all would close the case for impact.

THE RED DEVIL

Having concluded that the crater is likely to be in North America narrows the field, but it still leaves an impracticably large area to explore. When confronted with such a task, the intelligent geologist heads not for the field but for the library, there to scour the literature for reports of unusual K–T deposits and descriptions of circular geophysical patterns. The most diligent crater sleuth and literature searcher was Alan Hildebrand, a doctoral candidate at the University of Arizona who is now with the Geological Survey of Canada.[8] One report in particular caught his eye: the description by Florida International University's Florentin Maurrasse of a set of peculiar, late Cretaceous rocks exposed at the top of the Massif de la Selle on the southern peninsula of Haiti.[9] Maurrasse depicted a thick limestone sequence, the Beloc Formation, that contained a 50-cm layer that he thought was volcanic. In June 1989 Hildebrand visited Maurrasse and, as soon as he saw the Beloc samples, recognized them as altered tektites of impact rather than volcanic origin. Hildebrand then went to Haiti himself to collect from the Beloc Formation, and found the dual K–T layering that by that time had been described at many North American sites. In this case, however, the lower ejecta layer was about 25 times as thick as elsewhere and contained the largest tektites and shocked quartz grains ever found, suggesting that the Haitian site was close to the K–T target. Hildebrand and his colleagues estimated that ground zero was within 1,000 km of Haiti.

In May 1990, Hildebrand and William Boynton reported that the only crater candidate their literature search had turned up was a vaguely circular structure lying beneath 2 km to 3 km of younger sediment in the Caribbean Sea north of Colombia.[10] They acknowledged, however, that an impact at this seafloor site probably could not have provided the continental grains and rock fragments found in the boundary clay. Hildebrand and Boynton did note, almost as an afterthought, that in 1981, at the annual meeting of the Society of Exploration Geophysicists, geologists Glen Penfield and Antonio Camargo had reported circular magnetic and gravity anomalies from the northern Yucatán Peninsula and had speculated that buried there, beneath younger sedimentary rocks, might lie an impact crater.

By 1990, a peculiar kind of sedimentary rock deposit of K–T age had been found at several sites rimming the Gulf of Mexico. To the pro-impactors it appeared that these rocks were formed by giant waves, or tsunami, of the kind that a meteorite splashdown in the ocean would have produced. Since the Colombian basin site turned out to be the wrong age, the giant wave deposits helped to persuade

Hildebrand and Boynton that the Yucatán peninsula location identified by Penfield and Camargo was the more likely candidate. Soon after Hildebrand and Boynton announced in an article in *Natural History*[11] that it was indeed the K–T crater and claimed partial credit for the discovery: "In 1990, we, together with geophysicist Glen Penfield and other coworkers, identified a second candidate for the crater. It lies on the northern coast of Mexico's Yucatán Peninsula, north of the town of Mérida. The structure, which we named Chicxulub [pronounced Cheech-zhoo-loob] for the small village at its center, is buried by a half mile of sediments."*

In a letter to *Natural History* a few months later, reflecting the scientist's vital interest in priority, Penfield took exception to Hildebrand's claim, noting that *he* had "identified" the structure in 1978, and reminding readers of the words with which he and Camargo had closed their 1981 presentation: "We would like to note the proximity of this feature in time to the hypothetical Cretaceous–Tertiary boundary event responsible for the emplacement of iridium-enriched clays on a global scale and invite investigation of this feature in the light of the meteorite impact-climatic alteration hypothesis for the late Cretaceous extinctions."[12]

Whew! In 1981, only months after the Alvarez theory appeared, and in public at a scientific meeting, a prime candidate for the K–T crater had been identified, but no one had noticed. The Yucatán structure thus had to be rediscovered a decade later, after hundreds of person-years had been spent in the search. Poor timing may be a partial explanation: The meeting at which Penfield and Camargo presented their paper took place in the same week as the Snowbird I conference, to which the pro-impactors naturally were drawn. It turned out that the circular nature of the Chicxulub structure had been discovered in the 1950s by Petroleos Mexicanos (PEMEX) through the use of geophysical techniques. In the 1960s and 1970s, probing for possible oil-bearing structures, PEMEX drilled the structure and extracted rock cores. According to the account of Gerrit Verschuur, Penfield wrote to Walter Alvarez in 1980, right after he read the original Alvarez paper, to tell him that the crater might be located in the Yucatán, but he never heard back.[13] (At that time,

*Ever since the crater was named, uncertainty has prevailed over the proper translation of the Mayan word, *Chicxulub*. According to Mayan specialist George Bey of Millsaps College in Jackson, Mississippi, there are two acceptable meanings: "the place of the cuckold" and "the red devil." Since "red devil" is so evocative of the actual event, it seems the better choice.

however, Walter was focusing on the belief that the crater was located in an ocean basin and might not have taken a proposed continental site seriously.[14])

The failure of the geological community to jump at the Penfield-Camargo suggestion and save everyone a decade becomes even more poignant when we learn that a reporter for the *Houston Chronicle,* Carlos Byars, interviewed Penfield and Camargo in 1981 and wrote an article about their work.[15] In March 1982, an account of the Penfield-Camargo finding was published in *Sky and Telescope:* "Penfield . . . believes the feature, which lies within rocks dating to Late Cretaceous times, may be the scar from a collision with an asteroid roughly 10 km across."[16] How could it have been more clear that here was a lead that demanded to be followed up? And yet it was not. At least a few pro-impactors must have read that issue of *Sky and Telescope;* if so, they did not take the report seriously. Byars continued to attend meetings of pro-impactors to push the idea that the crater might lie underneath the Yucatán, but no one paid any attention. Perhaps it was necessary for the crater to be discovered, or rediscovered, not by a journalist or oil geologist, but by a bonafide member of the pro-impact research community.

Walter Alvarez argues that not finding the crater for 10 years "was actually a blessing" because it forced the pro-impactors to confront the repeated challenges from the Officer-Drake school.[17] Since one cannot rerun history and thereby learn what would have happened had he and others paid attention to Penfield and Camargo, it is hard to know whether he is right. In the long span of scientific history, a decade is not much, surely, and yet one can wonder what advances might have been made had the crater been confirmed in 1980 or 1981 and the next 10 years been spent differently.

TESTING CHICXULUB

The geophysical maps of the Chicxulub feature (Figure 14) showed it to have the form of a buried crater. The announcement of its discovery by a card-carrying pro-impactor, Alan Hildebrand, galvanized his colleagues into furious activity. Today, only a few years later, a vast amount of information has been assembled about the Chicxulub structure, more than enough to test the proposal that it is the K–T impact crater. Following the list of predictions given earlier, I will summarize the results of the many person-years of work on Chicxulub that have been crammed into the brief period from 1991 to 1997.

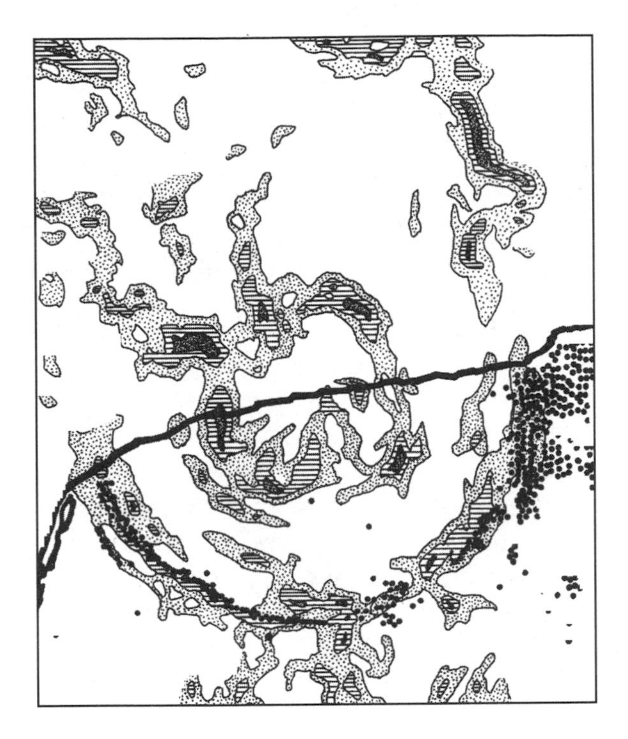

FIGURE 14 The cenote ring at Chicxulub (dark circles) superimposed on the gravity anomaly map. Note how the cenotes to the middle and left trace out part of a nearly perfect circle. [Photo courtesy of Alan Hildebrand and Geological Survey of Canada. For this and other images, see web page at http://dsaing.uqac.uquebec.ca/~mhiggins/MIAC/chicxulub.htm.]

SIZE AND SHAPE

Although the Chicxulub structure lies buried beneath a kilometer of younger sedimentary rocks, gravity anomaly maps clearly reveal it (see Figure 14). (Structural domes, which often contain petroleum deposits, also produce concentric geophysical patterns, thus explaining the decades-long interest that Chicxulub held for PEMEX.) The edge of the gravity anomaly indicates a structure at least 170 km in diameter, consistent with an impactor of 10 km in diameter. If Chicxulub is of impact origin, and if the outer perimeter of the gravity pattern represents the true outer rim of the crater, then that is its diameter. On the other hand, if the edge of the gravity pattern instead represents one of the inner concentric structures of a complex crater, for example, a collapsed terrace rim, then the structure might be larger, even much larger.

Still another possibility is that Chicxulub is not a complex crater of the kind shown in Figure 7 but something more: a giant, multiringed basin like the Orientale Basin on the moon. In any case, the original Alvarez calculation of 10 km was sufficiently imprecise that a greater radius for the crater is certainly possible. Virgil "Buck" Sharpton and his colleagues at the Lunar and Planetary Institute argue that Chicxulub may be as large as 300 km.[18] At 170 km, it would be one of the largest craters yet discovered on the earth; at 300 km, Chicxulub would be one of the largest impact basins formed in the inner solar system in the last 4 billion years. Since a 300-km crater would have released far more energy and been far more destructive to life than one just over half as wide, measuring the true size of Chicxulub is crucial to understanding the extinction. Drilling, geophysical studies, and topographic work of the kind being done by Kevin Pope and Adriana Ocampo of the Jet Propulsion Laboratory in Pasadena should soon resolve this dispute.[19]

The distribution of impact ejecta of various sorts can be used to learn more about how Chicxulub formed. Why, for example, is the shocked quartz in the North American sites only in the upper of the two impact layers, and why does more of it exist to the west than to the east of Chicxulub? Walter Alvarez, planetary-impact specialist Peter Schultz of Brown, and their colleagues are addressing both questions by studying the trajectory of the impactor and the mechanism of fireball creation and ejecta formation.

Note in the gravity anomaly map that the structure appears breached on the northwest side, an effect sometimes observed in craters on other bodies in the solar system. Peter Schultz believes that the K–T impactor was 10 km to 15 km in diameter and was descending toward the northwest at an angle of about 30 degrees from the horizon. According to Schultz, an impact from that angle and direction would explain the observed distribution pattern of shocked quartz, tektites, and iridium.

How important to the K–T event was the particular geology of ground zero in the Yucatán? Had the target area not been composed of limestones and sulfate deposits, would the extinction have been so great? Understanding the importance of the target rock types is critical to comparing Chicxulub with other potential impact-induced mass extinctions.

Topography

Kevin Pope, a consulting geologist now associated with the Jet Propulsion Laboratory, and two colleagues, came close to being the first to "rediscover" Chicxulub. He and Charles Duller of NASA were examining Landsat high-altitude satellite photographs, trying to determine the relationship between the ancient Mayan sites on the Yucatán Peninsula and the location of surface water deposits. They soon noticed a set of small ponds, which they later found went by the Mayan name of *cenotes,* that were arranged along the arc of an almost perfect circle. What could cause a set of small ponds to line up along the circumference of a circle? In an attempt to answer, Pope and Duller assembled PEMEX gravity and drill-core data and in 1988, during a conference in Mexico at which they first presented the cenote ring finding, showed the results to geologist Adriana Ocampo. She came to the brilliant conclusion that the combined gravity, core, and satellite data revealed a buried impact crater. Between 1989 and early 1990, the three worked on a paper to be submitted to *Science,* outlining their theory that a crater of K–T age was buried in the Yucatán. Before they were able to send it off, however, the 1990 *Science* paper by Hildebrand and Boynton appeared. Pope and his colleagues were astounded to learn that others were looking for a crater in the Yucatán. Pope then contacted Hildebrand, who sent preprints of a paper in which he named the structure Chicxulub. Pope, Ocampo, and Duller eventually saw their work published,[20] but by then priority for the rediscovery had gone to Hildebrand. Of course, the original discoverers were Penfield and Camargo.

As far as we know, the only process that can produce a circular ring with a diameter of 170 km is impact; no volcanic caldera is both so large and so perfectly circular. Pope, Ocampo, and colleagues interpret the cenote ring as having forming by postimpact collapse of the Yucatán limestones at the boundary between the fractured and unfractured zones that mark one of the impact rings, an effect commonly seen on other planets. If they are correct, the cenote ring is an inner circle, not the outer perimeter, and Chicxulub is much larger than 170 km. Later work on the morphology, topography, and soil types at the surface has led them to conclude that the crater is about 260 km in diameter.

Rock Types

In 1992, the indefatigable Officer and Drake weighed in.[21] Along with the late Arthur Meyerhoff, an American geologist who had been a consultant to PEMEX in the 1960s when the Chicxulub structure was drilled, they published an article in *GSA Today,* whose title, "Cretaceous–Tertiary Events and the Caribbean Caper," suggested that, far from having capitulated and accepted Chicxulub as the K–T impact crater, the authors intended instead to treat the notion as risible. Meyerhoff had been one of the most bitter opponents of plate tectonics and had made fossil identifications that, if correct, falsified the claim that Chicxulub was the K–T impact crater. He had a great deal at stake.

PEMEX's drilling of the Chicxulub structure in the 1960s and 1970s uncovered what their specialists at the time interpreted as a volcanic rock called andesite overlain by a thick sequence of Cretaceous sedimentary rocks, some of them brecciated. If this interpretation is correct, the claim that Chicxulub is the K–T crater would appear to be falsified on two counts: (1) it contains volcanic rather than melt rocks, and (2) a 10-km meteorite striking at the exact end of the Cretaceous would not leave a structure capped with Cretaceous rocks—the period ended with the impact. Thus for those who claimed that Chicxulub was the long-sought impact crater, two critical questions arose: (1) Is the igneous rock volcanic, or an impact melt? (2) What are the true ages and origins of the overlying sedimentary rocks? Only a detailed examination of the rocks could give the answers.

The required data could be obtained in two ways: by examining the older PEMEX drill cores and by drilling new holes. Unfortunately, the Coatzacoalcos, Mexico, warehouse in which the original PEMEX cores had been stored was destroyed in a fire, apparently leaving the notes that Meyerhoff had made on the specimens two decades earlier as the only extant reference source.[22] It eventually came to light, however, that a PEMEX employee had shipped a number of samples of the andesite to a colleague in New Orleans. Penfield then arranged to have some of those specimens sent to Hildebrand, whose associate David Kring immediately found that they contained shocked quartz. Meanwhile, new drilling efforts had begun to make new cores available for study; they showed that the Chicxulub rocks have relatively high concentrations of iridium and osmium, and reversed magnetism. Cores extracted from several holes drilled in the mid-1990s revealed the typical impact breccias, melt rocks, and structures of known impact craters. In particular, the Chicxulub sequence closely resembles that at Ries Crater.

AGE

Although its general stratigraphy and paleontology appeared to define the Chicxulub structure as Cretaceous, geologists argued about its exact age. Officer, Meyerhoff, and their colleagues said that since Cretaceous strata lay *above* it, the structure must be older than latest Cretaceous and therefore could not be the K–T impact crater. They based their conclusions on the earlier identification by Meyerhoff that the fossils in the breccia blocks above the melt rock were late Cretaceous. Subsequently, however, these fossils were redated and found to be from the Tertiary period. In any case, impact ejecta often contains blocks of older rocks that were excavated during a cratering event and blasted into the air, from whence they resettle, landing on top of younger ones in a way reminiscent of the upside-down-cake stratigraphy that Shoemaker found at Meteor Crater.[23] Ejecta from Ries Crater contains blocks of all sizes (some over 1 km) derived from rocks much older than the Miocene age of the crater. A single Chicxulub drill core could encounter such an older, out-of-place block on the way down and lead to the erroneous conclusion that, for that site, the K–T impact crater was falsified.

The first radiometric age report dated the Haitian tektites at 64.5 ± 0.1 million years, and for comparison, a feldspar from the K–T boundary at Hell Creek, Montana, source of *T. rex* skeletons, at 64.6 ± 0.2 million years.[24] These two ages do not quite overlap the 65-million-year age of the K–T boundary within their error bands. However, such bands reflect only the "intralaboratory" error, that is, they give the probable range within which the age would fall if measured again in the same laboratory. But the argon-argon method also requires reference to an interlaboratory standard, which can introduce small differences when different laboratories analyze the same sample, enough to bring the Haitian tektite ages into conformity with the K–T boundary age. Chris Hall of the University of Michigan and colleagues confirmed Izett's results in their own laboratory, obtaining an age of 64.75 million years for four separate Haitian tektites.[25] They noted that the ages of the tektites measured within their lab agreed so well that they "would make an excellent [argon-argon] standard"; for an isotope geochemist, this is the ultimate compliment. Izett's measurement represented the first time that a K–T impact product, if that is indeed what the Haitian tektites are, had been directly and absolutely dated.

Next came measurement of the age of the Chicxulub melt rock itself, and from two different laboratories. First to report were Carl Swisher and colleagues from the Institute of Human Origins at Berkeley.[26] They measured the age of the Chicxulub igneous rock and obtained 64.98 ± 0.05 million years, establishing the Chicxulub event as of exact K–T age. They also dated tektite glass from Haiti and glass embedded in rocks of K–T age at Arroyo el Mimbral in northeast Mexico (rocks that Smit and others believe were generated by the K–T event), and obtained almost exactly 65 million years for both. A few weeks later, Buck Sharpton of the Lunar and Planetary Institute and his co-workers reported an argon-argon age of 65.2 ± 0.4 million years for a different sample of the Chicxulub melt rock.[27] All of these measurements are consistent and show that the Chicxulub event, the Haitian tektites, and at least one K–T ejecta deposit, date to precisely 65 million years.

GEOCHEMISTRY

The Haitian glassy spherules are of two kinds: more abundant ones made of black glass, and rarer ones made of yellow glass. The chemical composition of the glasses indicates that they could have been derived from continental rocks of granitic composition, plus a minor component of limestone and clay.[28] This is consistent with an impact onto a continental shelf, where limestones, muds, and sulfur-bearing rocks are apt to be found, and which was the setting of the Chicxulub region during late Cretaceous time.

Imagine again the impact of a meteorite 10 km in diameter, but now focus on the target rocks and what happens to them. The temperature at ground zero instantly far exceeds the point at which limestone and sulfur-bearing rocks are converted into gases. These gases, along with vaporized meteorite and other target rocks, are lofted high into the atmosphere and distributed around the entire earth. Estimates are that billions of tons of both carbon dioxide and sulfur were injected into the atmosphere.

Not only did the chemical signatures of the Haitian glasses match those of the rocks from Chicxulub, so did the isotopes of oxygen, neodymium, and strontium. The isotopic measurements were made by Joel Blum and Page Chamberlain of Dartmouth, the university of Officer and Drake, showing that the geology department there was not monolithic in its view of the Alvarez theory.[29] Blum and Chamberlain found the Chicxulub igneous rock and the Haitian tektites to have identical isotopic ratios. The odds of this happening by chance are vanishingly small, and therefore not only are the two of the same age, as previously confirmed, they are linked by origin. However it was that the Chicxulub melt rock and the Haitian tektites formed, they come from the same source.

EJECTA DEPOSITS

Adriana Ocampo and Kevin Pope of the Jet Propulsion Lab, and Alfred Fischer (who moved from Princeton to the University of Southern California) have discovered the closest ejecta deposit to ground zero.[30] In a quarry on Albion Island in the Hondo River in Belize, 360 km from Chicxulub, at the top of the Cretaceous, they have come upon a double layer reminiscent of those found elsewhere around North and Central America, except that here some of the rock fragments are truly on a giant scale. The lower layer is about 1 m thick and contains abundant rounded spherules, 1 mm to 20 mm in size, composed of dolomite, a magnesiated limestone. Above it lies a 15-m layer containing broken fragments of a variety of rocks of Cretaceous age; some of the chunks are as big as a car.

Albion Island is important not only because it is the nearest Chicxulub deposit to ground zero, but because unlike all the others so far discovered, it appears to have been deposited above sea level, without the stirring and mixing effects produced by deposition in water.

The possibility that the impactor might have struck in the ocean and generated a tsunami, which in turn would have left behind characteristic deposits, was first suggested in 1985 by Smit and Romein.[31] They studied the K–T boundary in the Brazos River country of Texas, famed in American song and story, and found a rock known as turbidite (from the Latin *turbidus,* or disordered; with the same root as bioturbation). Turbidites are thought to be produced when sediment, jarred loose at the head of a submarine canyon by an undersea earthquake, cascades down as a submarine "landslide" and flows outward for hundreds and even thousands of miles. Turbidity currents produce layers of silt and sand several meters thick, with the coarsest material, which settles out first, at the bottom, grading upward to the finest sediment at the top. However, a giant wave, such as would be produced by a large meteorite striking near shore, would generate an earthquake which would also stir up sediments and produce a turbidite-like effect. It struck Smit and Romein as important that the only occurrence of such a rock type in the Brazos country was exactly at the critical K–T boundary, especially when that boundary also contained an iridium anomaly. Sedimentologist Joanne Bourgeois of the University of Washington concluded that the Brazos River turbidite was created by a tsunami 50 m to 100 m high.[32] Subsequent fieldwork identified many other examples of possible impact generated sedimentary deposits in the Gulf-Caribbean region. The one that by now is the most thoroughly studied is at Arroyo el Mimbral, one of several outcrops in northeast Mexico that expose the boundary layer.

"The significance of the Mimbral section lies in the combination of altered and unaltered glassy tektites, shocked minerals, anomalous iridium abundance, continental plant debris, and evidence for deep-sea disturbances and coarse-sediment transport precisely at the K–T boundary," according to Jan Smit and co-authors.[33] Charles Officer, Wolfgang Stinnesbeck of the University of Nuevo León, and Gerta Keller of Princeton were quick to disagree.[34] They too had studied Mimbral and "found no evidence of a nearby impact." Each piece of evidence presented by Smit and his co-authors, Stinnesbeck and company explained away. A key point in the disagreement was the position of the K–T boundary, which they placed above the level of Smit's uppermost unit, inevitably causing them to conclude that the rocks below it were older than, and therefore could not have been caused by, the K–T event. According to Smit, however, Stinnesbeck and colleagues placed the K–T boundary at the appearance of new Tertiary foraminifera. This is the way geologic boundaries have traditionally been set: at the level of first appearance of one or more abundant new species. However, a mass extinction caused by impact might delay the evolution of new species for tens of thousands or hundreds of thousands of years, until conditions were again hospitable. Therefore, Smit argued, to use the first appearance of new species to mark the boundary in the case of impact will always place the boundary too high, making it appear younger than it really is and by definition making the rocks immediately below it appear older than they really are. In contrast, ejecta and turbidites produced by impact would settle within a few hours or days of the event and would have exactly the same age as the impact.

On the contrary, Stinnesbeck and company argued, the deposits at Mimbral and elsewhere formed through normal geologic processes, possibly as coastal sediments slumped into deeper water. In that view, the Gulf of Mexico K–T sections, including Mimbral, were not impact-generated, but are of pre-K–T boundary age and were probably deposited by turbidite or gravity flows.[35] Thus the debate turns on whether the Gulf K–T deposits, as Alfred Fischer put it in an address at the Snowbird III conference, "formed in 100,000 seconds [1.15 days] or 100,000 years."[36]

The controversy resembles the one over the sharpness of the iridium peak at Gubbio. In both cases, experts viewing the same evidence come to opposite conclusions and debate the matter at length in the literature. But having seen the example of the blind test at Gubbio, the solution to the Mimbral controversy became obvious, though it needed a modification: Since the outcrop could not be brought to the geologists, the geologists had to be brought to the outcrop. Accordingly, a field trip to Mimbral was organized at the time of the 1994 Snowbird III conference, held not in Utah but in Houston. In the party were four former presidents of the Society for Sedimentary Geologists, including Robert Dott of the University of Wisconsin, regarded by many as the dean of American sedimentologists. According to Richard Kerr, the *Science* magazine reporter who has made the impact controversy a specialty, Dott spoke for these experts on sedimentary rocks: "We were impressed with the evidence that this sequence was very rapidly deposited. [It must have taken] closer to 100,000 seconds than 100,000 years."[37]

According to Kerr, the assembled sedimentologists concluded that the most likely cause of what they had seen was an impact-generated tsunami. Keller, ignoring Mark Twain's advice never to get into a contest with a man who buys ink by the barrel, took vigorous exception to Kerr's account. In her rebuttal, she claimed that "The impact tsunami scenario did not win the day. . . . Sedimentologists generally disagreed with Smit's model of tsunami wave deposition."[38] Kerr replied, "Of the five sedimentologists on the trip other than Lowe that I interviewed for the story, four of them agreed that the deposit is consistent with waves from an impact and that no proposed alternative can reasonably explain the deposit."[39]

Predictions Met

Officer followed up his 1992 paper (with Drake and Meyerhoff) with one in the journal *Geology* (with Dartmouth colleague J. B. Lyons and Meyerhoff) and, later, with a section in his book with Page, *The Great Dinosaur Extinction Controversy*.[40,41] In the book, after setting up the anti-impact position by citing the various (to him at least, successful) challenges to the Alvarez theory that Charles Officer has championed through the years, the two authors come at last to the vexing issue of the crater itself. But for them, Chicxulub is not vexing at all; they merely rebury it: "The impactors [pro-impact scientists] would have *something* of a case if they would point to a massive impact crater . . . dating to the proper time in the geological record. They have searched far and wide around the world for evidence of even one such crater, but sadly for them, they have come up wanting. . . . One of the things that did not happen at the K–T boundary was impact by a gigantic meteorite."[42]

How, in the face of all the evidence just reviewed, can two authors come to such an opposite conclusion from almost everyone else who has studied the Chicxulub crater? Here is how Officer and Page managed it:

- Their book appeared in July 1996, allowing plenty of time for them to incorporate the results of the new drilling tests that by 1994 had begun to be reported at scientific meetings and in abstracts. Yet Officer and Page base their conclusions on only two sources: a report from 1975 and Meyerhoff's three-decades-old notes.

- They do not mention the strikingly concentric gravity patterns, the cenotes, or the import of the size of the structure (at 170 km to 300 km, perhaps the largest on the earth).

- The older paleontologic interpretations dated the Chicxulub structure at 80 million to 90 million years, far older than K–T time. Officer and Page mention the modern radiometric age measurements, but say only that they "give values ranging from

58.2 to 65.6 million years...in accord with what would be expected from samples...with loss of argon content."[43] They compare this with the revision of Manson's age (although there the more recent measurement gave an older, not a younger age). To the unwary, their discussion leaves the distinct impression that the most recently obtained ages at Chicxulub are suspect and that the original ones stand unchallenged. Officer and Page do not reveal that two different laboratories conducted the modern age measurements on the Chicxulub igneous rock, that they used the highly precise argon-argon method, and that both gave precisely the same result—exactly 65.0 million years. It is true that in one of the studies, a few of the samples gave ages as low as 58.2 million years, which the original authors attributed to alteration, but most from that study gave 65.0 million years. Since argon loss causes ages to be younger than they really are, not older, the older ages are the more reliable. As testimony, in the other of the two dating studies, three samples of the Chicxulub melt rock gave 64.94, 65.00, and 64.97 million years.

- Although they list among their references the paper by Dartmouth scientists Blum and Chamberlain, who used isotopic ratios to establish a genetic link between the Chicxulub melt rock and the Haitian tektites, Officer and Page never actually mention this result in their text.

- They do not reveal that the Chicxulub igneous rock has anomalously high iridium levels. They acknowledge that it does contain shocked minerals, but pass them off as "of the volcanic/tectonic type."[44] (The original authors, however, clearly stated that the shocked minerals show the multiple deformation planes indicative of impact, features that have never been found in volcanic rocks.[45]) They do not mention that the Chicxulub melt rock is reversely magnetized, consistent with (but not proof of) a K–T age.

- Although in an earlier section they discuss the visit of the sedimentologists to Mimbral, in their section on the crater search they never mention the conclusion that the sedimentologists reached: that the Mimbral sediments "were deposited on short time scales (more likely 100,000 seconds than 100,000 years)."[46]

Chicxulub has met each of a reasonable set of predictions for the impact crater, and then some. First, the concentric gravity patterns and its huge size show that Chicxulub is not a volcanic feature but an impact crater. Second, with its breccias and a melt rock that is reversely magnetized and enriched in iridium and shocked minerals, it has the features expected of the K–T crater. Third, Chicxulub formed at precisely the time of the K–T boundary. Fourth, its age and isotopic geochemistry link it conclusively to the unusual Haitian tektites. Finally, turbidite-like K–T deposits enriched in iridium, spherules, and spinels, and believed by most experts to have been laid down in a day or so, encircle Chicxulub. In sum, if this structure is not the K–T crater, it is hard to imagine what would ever qualify. One could describe this result using the tiresome metaphor of "smoking gun," but in this case it is not really apt. When the perpetrator is long gone, merely finding a smoking gun is not enough—you need to know whose fingerprints are on it.

The Zircon Fingerprint

Although the list of features of Chicxulub is long and closely matches those to be expected of the K–T impact crater, each individual piece of evidence is circumstantial. To be precise, what has been demonstrated is that on the Yucatán Peninsula, buried under a half-mile of sedimentary rock, lies a crater of K–T age that is somehow linked to the unique Haitian tektites. What has not been demonstrated is that this crater is the parent of the K–T boundary clay around the world. It is highly likely, to be sure, but not proven. But surely circumstantial evidence is the best we can hope to find for an event buried so deep and so far in the past.

Not so. A new kind of evidence, based on the fractured zircon that caught my eye on the cover of *Nature,* offers proof of a genetic link between the crater and the boundary clay and completely rules out volcanism. To understand this evidence, we need to go a bit deeper into the way radioactive parent-daughter pairs are used to measure rock ages. The principle is simple: Atoms of some elements spontaneously emit subatomic particles, such as neutrons and protons, and in so doing change into atoms of other elements. Each decay occurs at a known rate, called the half-life, which is the time it takes for one-half of any original number of parent atoms to convert to daughter atoms. If we know how much of the parent and how much of the daughter are present in a sample today, and we know how rapidly the parent changes into the daughter (the half-life), we can calculate how long it took for the original amount of parent to decay to that amount of daughter. As an analogy, imagine that you enter a room at exactly 9:00 A.M. and find an hourglass standing on a table, with three-quarters of its sand in the bottom cone. You would quickly conclude that the sand has been flowing for 45 minutes and therefore that the hourglass had been turned over at 8:15 A.M. Radiometric dating is similar but adds two wrinkles. First, imagine that as the grains pass through the constriction between the upper and lower cones of the hourglass, they change color, analogous to one element turning into another. Second, imagine that the constriction is adjustable and is tightened a little more as each grain of sand falls through. The speed with which the grains fall from top to bottom is then related to how many grains remain in the upper half: The fewer that are left, the more slowly they fall through. You could no longer figure out in your head when the hourglass had been turned over, but if the tightening followed certain rules, a simple mathematical formula would do the trick.

Of course, in practice radiometric dating is not so simple. Sometimes atoms of the daughter element, inherited from some ancestral rock, were already present when the decay clock started to run, thus causing the rock being dated to appear older than it is. In other cases, parent or daughter atoms are gained or lost after the decay clock has started to run, throwing off the calculation. Daughter loss is commonly caused by heat, which expands and opens crystal structures and allows the loosely bonded daughter atoms to escape (this is a particular problem with argon atoms when using the older potassium-argon method). If all the daughter atoms are lost, the radiometric clock is set back to zero and the time subsequently measured is not the true, original age of the specimen, but rather the time that has elapsed since it was heated. However, geochemists know how to tell when each of these problems has arisen and usually can correct for them.

The decay of uranium into lead is unique among the geologically useful parent-daughter pairs because two isotopes of uranium, each with its own half-life, decay into two isotopes of lead: U 238 decays to Pb 206 and U 235 decays to Pb 207. Thus two uranium-lead clocks keep time simultaneously. If one plots the uranium-lead isotopic ratios of samples that have suffered no lead loss, they lie along a special curve called Concordia (after the goddess of agreement), shown in Figure 15. If several samples of the same rock or mineral plot at the same point on Concordia, we know that the material has not lost uranium or lead and that the date obtained is its true original age. Because uranium is present at measurable levels in a variety of rocks and minerals, the method has wide applicability.

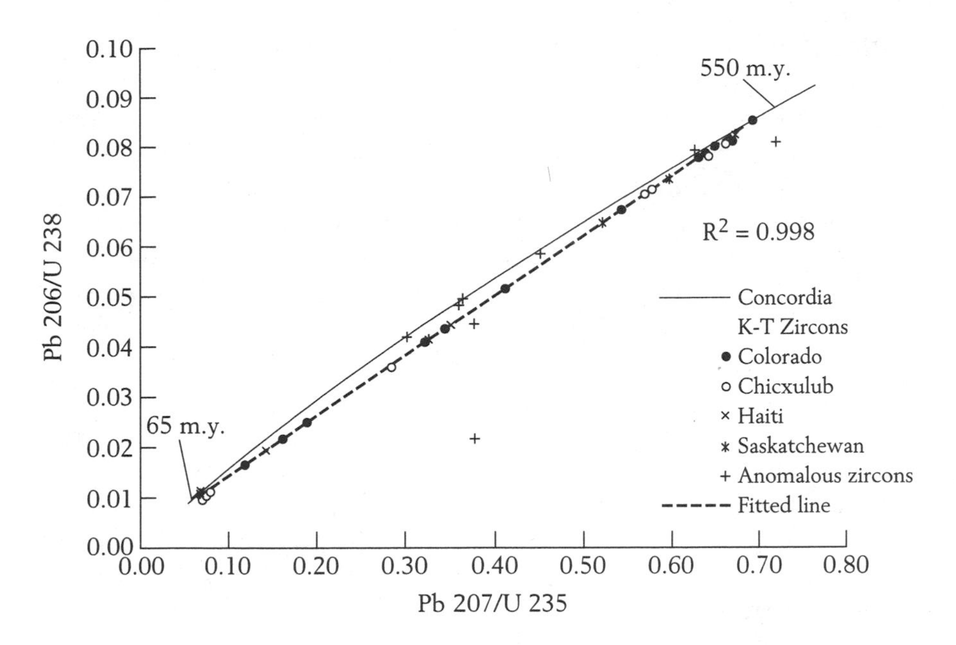

FIGURE 15 Composite diagram showing a section of Concordia (the smooth curve) and the position of K–T zircons from Chicxulub, Haiti, Colorado, and Saskatchewan. The fit of the lead-loss line is nearly perfect. A few points that appear to have a different history are also shown. [Data from Krogh and colleagues; recalculated by the author.]

Sometimes, however, the uranium-lead ratios of a suite of related specimens plot not on Concordia, but on a straight line that intersects it twice, like a chord to an arc. The mathematics of uranium-lead decay reveal why: When a rock or set of minerals has been altered and lead has escaped, samples with different degrees of lead loss plot along a line that intersects Concordia at two points. The upper, older intercept of the straight line and Concordia gives the original age of the rock; the lower, younger intercept gives the time at which the rock lost lead. When measured ratios plot neither on Concordia nor along a straight line, we know that the geologic history of the samples is more complicated—they may have passed through more than one heating event or they may have lost variable amounts of both uranium and lead. Such results give little or no useful information, and the geochronologist tries again with different samples. But when several samples do plot on a straight line that intersects Concordia twice, we know that the material has passed through only a single lead loss episode and that we have measured both its original age and the time at which it lost lead. Uranium-lead dating thus provides two essential pieces of information and is one of the most powerful tools in the geologist's kit.

The ideal substance for uranium-lead dating would be one that contained no original lead but enough uranium to have produced measurable amounts of radiogenic lead (lead derived from uranium decay) over periods of geologic time. It should occur in a variety of rock types. While we are at it, why not ask for a mineral that is so hard and chemically inert that it survives weathering, erosion, and the heat and pressure of metamorphism? Believe it or not, exactly such a mineral exists: zircon. When zircon crystallizes, it contains uranium but no lead, thus eliminating the problem of original daughter atoms. A minor by-product in granitic rocks, zircon sometimes grows large enough to form a gemstone (a fact well known to viewers of home-shopping networks). Because it survives erosion and every geologic process known except complete remelting, zircon winds up in a wide variety of rocks and looms much larger in understanding earth history than its infrequent occurrence would suggest.

The application of zircon dating to the K–T boundary problem now begins to become clear. If the rocks that existed at ground zero contained zircons, which many continental rocks do at least in small amounts, these zircons might have been shocked, heated, and had their clocks at least partially reset. The fireball cloud might have lofted them high and distributed them over thousands of miles. If suites of such zircons show up in the K–T boundary clay, their uranium and lead isotopes might have retained both the original age of the target and the time of the impact—65 million years. They would then fall along a straight line that intersected Concordia at 65 million years and at some older age defined by the true age of the target rocks. But surely this is too much too expect.

Bruce Bohor of the U.S. Geological Survey and Tom Krogh, who now operates one of the world's most sophisticated lead-dating laboratories, at the Royal Ontario Museum in Toronto, examined samples from the upper K–T layer at the Raton Basin in Colorado and discovered zircons with the same multiple shock deformation lamellae that characterize impacted quartz.[47] Shocked zircon had never before been seen. Krogh and his Royal Ontario Museum colleague Sandra Kamo set out to measure the uranium and lead isotopic ratios of these zircons, but ran into two difficulties. First, although the analysis had to be done grain by grain, the individual zircons weighed only from 1 millionth to 3 millionths of a gram and could not even be seen with the naked eye, making them hard to handle, to say the least. Second, the zircons contained only between 5 picograms and 200 picograms (a picogram is a trillionth of a gram, or 10^{-12} gram) of radiogenic lead, almost too little to be measured. This made analysis especially difficult because, as with iridium, it is almost impossible to rid a laboratory of the effects of contamination from environmental lead. To overcome these two problems, Krogh and Kamo invented new methods of lead analysis, in the process reducing their laboratory lead background (the amount of environmental lead contamination that cannot be eliminated) to the lowest of any lab: 2 picograms of lead per experiment.

Using an electron microscope, they found that they could arrange the tiny Raton Basin zircons visually in a series with unshocked specimens at one end and increasingly shocked and finally granular zircons (whose crystal structures had been completely destroyed) at the other. On the Concordia diagram, the uranium-lead isotopic ratios of these zircons plot along a nearly perfect straight line, showing that they came from a single target and experienced a single episode of lead loss. The line intersects Concordia once at 550 ± 10 million years, which must be the original age of the zircons, and again at 65.5 ± 3 million years, which must be the time of lead loss. The more shocked the zircons, the farther down the line they plot, closer and closer to the lower, 65-million-year intercept. This point is absolutely critical: Zircons that originated in a volcanic eruption 65 million years ago would have crystallized at that time. They would possess, and would display, an original age of 65 million years—not 550 million years. Subsequent lead loss would make them appear younger than 65 million years, not older. The least altered and unaltered zircons would give the true age of 65 million years. But just the opposite is the case for the Raton Basin zircons: The unshocked and least shocked zircons give the oldest ages, while the most shocked, including those that are completely shocked, give the youngest, approaching and in extreme cases reaching 65 million years. Because 65-million-year-old zircons could not produce this result, neither the zircons nor the clay itself could have come from a 65-million-year-old volcanic eruption.

Krogh and his group went on to take the next logical step: They analyzed zircons from both the Chicxulub breccia and from the Haitian Beloc Formation.[48] Plotting the Chicxulub, Haitian, and Raton Basin zircons on the same Concordia diagram, they found that 18 of 36 fell on a straight line that intersected Concordia at 545 ± 5 million and 65 ± 3 million years, "as though they had come from a single sample."[49] Finally, the group studied zircons from the K–T site most distant from Chicxulub, in South Central Saskatchewan, 3,500 km away.[50] They found an upper intercept age of 548 ± 6 million years and a lower of 59 ± 10 million years, the same within the analytical precision as the results from the other three sites.

In all, Krogh and his colleagues studied 43 K–T zircons. A few seemed to point to an age of about 418 million years for the parent rock; several others scatter randomly when plotted on the Concordia diagram, indicating they have had a more complex history, perhaps having lost lead twice. But as shown in Figure 15, of the 43 zircons from all four sites, 30 fall exactly on a straight line (the 418-million-year-old zircons are omitted from the diagram). These 30 zircons, found at four sites separated by thousands of kilometers and representing three completely different geologic settings—a Chicxulub breccia, Haitian tektites, and K–T boundary clays from two locations, one 3,500 km from the Yucatán—plot precisely on a single straight line with a coefficient of correlation, the statistician's test of "goodness of fit," of 0.998. (When all 43 zircons are included, even those that obviously have a more complex history, the correlation coefficient is still a remarkably high 0.985.) If one were to collect and analyze 30 zircons from a single rock unit, their fit could be no more perfect. Even though scattered over 3,500 km, these are the same zircons.

One of the most surprising results of Krogh's work, after one gets used to the near perfection of the fit, is that so many of the zircons have the same original age. Since we know that the impact that formed Chicxulub excavated a crater some 20 km deep, a huge slice of crustal rocks, with diverse ages and compositions, should have been caught up in the ejecta. Yet most of the zircons give the same 550-million-year upper age. Krogh and his co-workers speculate that the upper few kilometers at the Yucatán ground zero may have been made up of zircon-free limestone, so that most of the zircons came from a single underlying, zircon-bearing layer.

The remarkable sleuthing of Krogh and his colleagues has to rank as one of the great analytical triumphs of modern geochemistry (though Officer and Page do not cite them in their 1996 book or once mention zircon). Here is what the zircons tell us:

1. The K–T boundary clay was not formed by volcanism. Had it been, none of the K–T zircon ages would exceed 65 million years. Furthermore, volcanic zircons are angular and unshocked, not the opposite.
2. At least in the Western Hemisphere, the clay had a single source crater rather than having been derived from multiple impacts, as some had suggested in the late 1980s.

3. The K–T ejecta deposits from Haiti, Colorado, and Saskatchewan each came from the same target rock, of age 545 million to 550 million years, and each was shocked at exactly the same time: 65 million years ago.
4. Since rocks 545 million to 550 million years in age are rare in North America, and since the Chicxulub zircons themselves give both that upper age and the 65-million-year lower age, the K–T ejecta in Haiti, Colorado, and Saskatchewan almost certainly came from the Chicxulub structure.

Thus the K–T zircons provide direct, noncircumstantial evidence that Chicxulub is the K–T impact crater. This is no longer a fascinating speculation, but closely approaches the status of observational fact. Today, those who doubt that Chicxulub is the long-sought crater can be counted on one's fingers. Yes, a giant meteorite did strike the earth at the end of the Cretaceous. But did it cause the K–T mass extinction and the death of the dinosaurs?

Part III

Did an Impact Cause the K–T Mass Extinction?

Chapter 8

Clues from the Fossil Record

With respect to the apparently sudden extermination of whole families or orders . . . we must remember what has already been said on the probable wide intervals of time between our consecutive formations; and in these intervals there may have been much slow extermination.[1]

Charles Darwin

What Extinction?

Craters are physical features. Even one hidden under a kilometer of rock can be discovered using geophysical techniques, then drilled, and samples brought back to the surface and studied in the laboratory. As more and more craters have been discovered on earth using such techniques, and as the evidence that Chicxulub is the K–T impact crater has accumulated, more and more geologists have come to agree that a giant impact ended the Cretaceous. Specialists have calculated that the crash released the energy equivalent of 7 billion bombs the size of the one dropped on Hiroshima, and produced the loudest noise heard and the brightest light seen in the inner solar system in the last 600 million years. Such an event, like the passage of billions of years, is far beyond our experience and our ability to comprehend. Surely nothing could more clearly refute Hutton's maxim, "The present is the key to the past." Confined as we are to the present, it has taken geologists nearly 200 years to discover that large meteorites have struck the earth and that terrestrial craters—many of them—exist. That recognition leads to a new question: What are the consequences of a giant impact for living creatures, such as those that inhabited our planet 65 million years ago? On that point, far less agreement exists. There is consensus, however, that the answer is to be found in the fossil record.

The question of the effect of impact on life is surely important, for if the collision that left the Chicxulub crater behind did *not* cause the extinction of the 70 percent of species that perished at the end of the Cretaceous, the Alvarez theory would remain merely a scientific curiosity. Yes, objects from space do strike the earth now and then, but even one the size of a large mountain does little harm to life (or to geological orthodoxy). If on the other hand, impact did cause the extinction, then paleontology, geology, and biology would never be the same. Our conception of the role of chance in the cosmos, our view of life and its evolution, our understanding of our own place—each would be irrevocably altered.

The Alvarez team left no doubt that they believed that the impact caused the mass extinction. They might have called their 1980 article in *Science* "Evidence for Impact at the Cretaceous–Tertiary Boundary" and waited until the case for impact was strongly corroborated before going on to connect it to mass extinction. Instead they gave it the provocative title, "Extraterrestrial Cause for the Cretaceous–Tertiary Mass Extinction" and set out to show both that impact had occurred *and* that it had caused the K–T mass extinction.

The original paper was unusually long for *Science*, indicating that Luis's Berkeley protégé from long ago, editor Philip Abelson, understood that the new theory might be of more interest than most. However, almost all of the article's 13 pages were devoted to describing the iridium measurements and other chemical tests; the biological consequences of impact covered only half a page. This was undoubtedly because the Alvarez team, though long on scientific talent, was demonstrably short on knowledge of paleontology. Luis was a physicist; Asaro and Michel were chemists; Walter was a geologist, but not a paleontologist. In retrospect, since they did not have paleontological credentials, it was both proper and good strategy for the Alvarezes to introduce their theory but to leave to others the task of testing it against the facts. Walter knew that Wegener's theory of continental drift had languished for decades in part because in seeking a mechanism to explain why the continents had drifted, he strayed outside his field into the territory of the geophysicists, who immediately pronounced drift impossible, thus putting an end to the matter for half a century.

For the first few years after its appearance, paleontologists did not believe that they needed to take the Alvarez proposal seriously. Even if impact were strongly corroborated—even if the crater were found—that would not necessarily mean that the impact had caused the mass extinction. More importantly, paleontologists believed that for the all-important dinosaurs, the question had already been answered: Evidence that they had collected over more than a century, and especially in the Montana dinosaur beds in the 1970s and early 1980s, they interpreted to show that the dinosaurs had gone extinct gradually, not instantaneously. This point of view infuriated the already irascible Luis Alvarez: "I simply do not understand why some paleontologists—who are really the people that told us all about the extinctions and without whose efforts we would never have seen any dinosaurs in museums—now seem to deny that there ever was a catastrophic extinction. When we come along and say, 'Here is how we think the extinction took place,' some of them say, 'What extinction? We don't think there was any sudden extinction at all. The dinosaurs just died away for reasons unconnected with your asteroid.'"[2]

RETURN OF THE PTERODACTYL?

The founders of geology and biology were not much interested in extinction. Lyell thought that extinction was so impermanent that the vanished pterodactyl might return to flit through a forest once again primeval. Darwin thought that gradual change was the essence of natural selection: "Species and groups of species gradually disappear, one after another, first from one spot, then from another, and finally from the world."[3] "Extinction and natural selection . . . go hand in hand."[4] He did occasionally make exceptions: "In some cases . . . the extermination of whole groups of beings, as of the ammonites towards the close of the secondary period, has been wonderfully sudden."[5] (Scientists of Darwin's day thought there were four main geologic periods: primary, secondary, tertiary, and quaternary.)

Extinction was simply the natural end of every species and therefore unremarkable. Biologists were much more interested in speciation, the process by which an evolutionary lineage divides, giving rise to two species where only one existed. Reflecting the increased interest in extinction since the Alvarez theory appeared, paleontologist David Raup of the University of Chicago and the Field Museum has written a fine book on the subject, *Extinction: Bad Genes or Bad Luck?*[6] Raup's most fundamental conclusion about mass extinction, drawn from a lifetime of study, is that, because species typically are well adapted to the normal vicissitudes of life, "for geographically widespread species, extinction is likely only if the killing stress is one so rare as to be beyond the experience of the species, and thus outside the reach of natural selection."[7] His conclusion is key to understanding the role of meteorite impact in earth history.

The core concept of natural selection is that species continuously adapt to their environment. This means that organisms tend to be well suited to the normal stresses they encounter, even those that occur on a time scale of thousands or hundreds of thousands of years. (Note that Raup said that the stress has to be unfamiliar not only to *individuals* but to entire *species*.) Normal environmental changes, such as a gradual lowering or raising of sea level, or a gradual alteration in climate, cannot by themselves cause a mass extinction—they allow species time to adapt or to migrate to more favorable climes. Some species, it is true, will be unable to do either and will become extinct, but those few do not a mass extinction make. This is more than theory. Another Chicago paleontologist, David Jablonski, examined the fossil record from major mass extinctions in the geologic record to see if they correlated in any way with known changes in sea level, global climate, and mountain building; he found that they did not.[8] Of course, catastrophic events such as floods, earthquakes, and volcanic eruptions can and do kill many individuals and, on rare occasions, even species (if they occur in a limited geographical area), but these are not global or even continental in their reach. For species, the opposite of the old saying, "What you don't know can't hurt you" is true: It is what a species does know that can't hurt it.

But suppose that Raup is wrong and that global cooling can cause a mass extinction. Then the evidence should be readily at hand, for the earth has just suffered a succession of ice ages, the last one ending only some 15,000 years ago. Temperatures fell so far that huge ice sheets advanced thousands of miles, covering, for example, most of the northern half of the United States. Many of the large mammal species (of perennial fascination to *Homo sapiens*) did become extinct during the last ice age, but the overall extinction rate was far below that of the five major mass extinctions and barely makes it onto a chart of extinction intensity (Figure 16). Admittedly, scientists

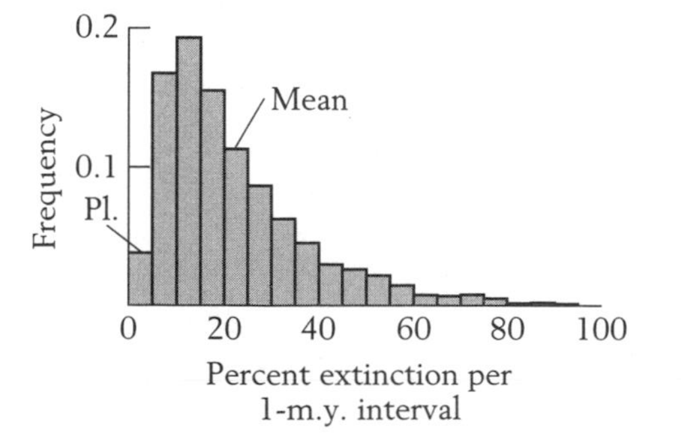

FIGURE 16 Variation in intensity of extinction for the last 600 million years broken up into 1-million-year intervals. [After David Raup.] The Big Five are out on the right tail; Pl refers to the Pleistocene extinction rate, which is far below that of a mass extinction, even though climate and sea level changed drastically.[9]

have not figured out how the ice ages spared so many species, though perhaps the ice advanced slowly enough to allow some to migrate to warmer climes, while others may have been preadapted for cold or survived in ecological refuges. Some think that the arrival of skilled aboriginal hunters on a virginal North American continent has much to do with the Pleistocene mammal extinction, but many disagree, pointing to the millennia of coexistence of humans and large mammals in Africa.

If we accept Raup's conclusion that species living over a wide area can be killed off only by stresses with which they are unfamiliar, and that those stresses must occur too rapidly for migration or adaptation, it follows that rare, sudden, and global catastrophes must also exist—otherwise there is no way to explain the several mass extinctions that mark the geologic record. This is worth repeating: To accept that global mass extinctions have occurred is also to accept that global catastrophes have occurred, a conclusion that is the antithesis of strict uniformitarianism.

The five largest mass extinctions in terms of percent of species killed—the Big Five—are shown in Table 2. Note that the record for intensity is held not by the K–T but by the end-Permian extinction.

Until recently, paleontologists believed that extinction came in two forms: a regular, low-level, background extinction, and the much more destructive mass extinctions. Over the last few years, they have had the benefit of databases meticulously compiled by such paleontologists as John Sepkoski, also of the University of Chicago. He and Raup studied the fossil record of over 17,000 extinct genera of marine animals, and several times that many species. Their database shows that the mean duration of a genus is about 20 million years and that of a species is about 4 million years.

TABLE 2

The Big Five Mass Extinctions

Extinction episode	Age (million years before present)	Estimated species extinction %
Cretaceous–Tertiary	65	70
Triassic–Jurassic	~202	76
Permian–Triassic	~250	96
Late Devonian	367	82
Ordovician–Silurian	~438	85

The database can also be used to help answer a question over which paleontologists have puzzled: Are the giant extinctions fundamentally different from the background extinctions, or do they merely represent the extreme end of a continuum? Figure 16 helps us decide. This chart was constructed by dividing the 600 million years since life began to flourish, at the beginning of the Cambrian period, into intervals of 1-million-year duration, and computing for each the number of species still alive at the end as a percentage of those alive at the beginning. The mean is a 25 percent extinction rate per million years (on the average, of 100 species alive at the beginning of a 1-million-year period, 75 were still alive at the end). The Big Five lie off on the right tail, but there is no break between them and the lesser extinctions—the distribution appears continuous. Background and mass extinctions therefore do not seem to be qualitatively different, but rather to grade imperceptibly into each other. If all extinctions had a common cause, but one that operated at different intensities at different times, this is the pattern we would expect. We cannot say, however, that some combination of extinctions with different causes might not give the same result.

The great extinctions reached diverse organisms in almost every ecological niche. The K–T extinction wiped out animals as unlike as microscopic foraminifera, intricately coiled ammonites, land plants, and dinosaurs—from the tiniest creatures of the sea to the largest denizens of the mountain slopes. Obviously, such different organisms, in such completely different environmental settings, did not compete in Darwin's sense. Most species that died appear to have been as successful as those that survived. Before their fall, there would have been no reason to predict that they would be the ones to go, yet go they did. The converse is also true. In most cases it is impossible to say why the species that survived did so; certainly it was not because they were more "fit." Thus, Raup concludes, evolution and survival may be more matters of chance than fitness, of good luck than good genes. In his view and that of the Alvarezes, the dinosaurs, and the others that joined them in disappearing at the end of the Cretaceous, were more than anything unlucky enough to be in the wrong place at the wrong time.

THE FOSSIL RECORD

When the Alvarez theory broke upon the world, most paleontologists were quite confident that it could immediately be judged by the weight of more than a century's worth of fossil evidence and rejected out of hand. But they were wrong. An obstinate set of problems makes interpreting fossil data so difficult that testing the extinction half of the Alvarez theory proved much harder than anyone could have anticipated. In the testing, however, so much was learned that some paleontologists believe that their field is undergoing a procedural if not a scientific revolution. For the first time, large collections are being established with the specific aim of testing whether extinction near a major geologic boundary was gradual or sudden, and whether species that have always been thought to have gone extinct at a boundary truly did so.

In order to understand how the second half of the Alvarez theory was tested, it is important first to recognize some of the problems inherent in trying to read the fossil record. Common sense tells us that to corroborate the extinction half of the theory, we need to find two kinds of evidence: (a) that prior to the K–T boundary, most species were not already going extinct for some other reason, and (b) that the dinosaurs and others did not survive the K–T impact—that their remains do not lie above the iridium layer. To test both predictions, geologists needed to be able to pin down the exact point in a sequence of rocks at which the extinction of a particular species occurred. Can that be done?

GAPS

Darwin recognized, as noted in the epigraph that opens this chapter, one insuperable problem with interpreting the fossil evidence: Erosion has caused the geologic record to be riddled with missing rock units. As shown in Figure 17, a missing unit can lead to the false conclusion that a fossil species became extinct earlier and more suddenly than it actually did. Therefore, unless there is independent evidence that a geologic section contains no gaps, an apparently sudden extinction cannot be taken at face value. This works against the

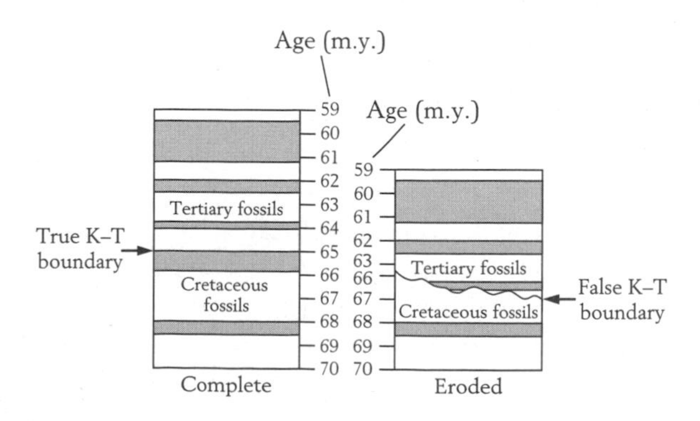

FIGURE 17 Erosion distorts the geologic record. [After Raup.[10]]

pro-impactors, who hope to use a pattern of sudden extinctions as evidence of catastrophe; they must first establish that the section they are studying is free of gaps.

Sometimes gaps are obvious (there's a spot in the Shoshone Gorge in Wyoming where you can put your finger on one representing 2 billion years, for example). But especially in rocks formed in the deep sea—limestones and mudstones, for example—gaps may be nearly impossible to detect. Due to the intense study the K–T part of the geologic column has received since the Alvarez discovery broke, more gaps have been discovered there than had ever been imagined.

Some geologists studying the K–T boundary found what they thought was a pattern of extinction intermediate between catastrophic and gradual. In this so-called stepwise extinction, species appeared to disappear in sets, one after the other, as the K–T boundary was approached. What could explain stepwise extinction? Some proposed that a cluster of meteorites had fallen one after another, each wreaking its own bit of havoc and each causing an extinction. A large comet might have broken into pieces that then went into orbit, and these pieces might subsequently have fallen to earth one after another, like the Shoemaker–Levy 9 "string-of-pearls," though over a much longer period of time. If stretched a bit, this idea could accommodate everyone: Impact had occurred, not once but several times, satisfying the pro-impactors; the sequence of impacts gave rise to a kind of gradual extinction, pleasing the paleontologists. Not a bang, but something more than a whimper. And because volcanism tends to occur in pulses over geologic periods of time, stepwise extinctions also had a natural appeal for the volcanists. Everyone could be happy! Distinguished scientists from various sides of the debate came together to co-author papers proposing multiple impacts and stepwise extinctions. Gaps can also produce a false stepwise extinction pattern, however. As the evidence has accumulated favoring a single impact, interest in the stepwise variation on the original Alvarez theory has waned.

MIGRATION AND DISSOLUTION

Another process that makes interpreting the fossil record difficult is that species, rather than going extinct, sometimes simply abandon an area in favor of another one nearby. If today we were to search only locally, we might mistakenly conclude that the species had become extinct. Yet a broader search finds it on a nearby island or in an adjacent region. It did not expire, it moved. Migration causes us to underestimate the true ranges of species in the fossil record and to think they became extinct before they actually did. (In Chapter 9 we see a good example in the ammonites.)

Fossil material is usually composed of carbonate or phosphate, chemicals that can dissolve in groundwater. Thus even though an organism becomes fossilized, its remains may later be dissolved away and disappear. This causes us to underestimate the true range of a fossil species, or in the extreme, to miss it entirely.

POOR PRESERVATION AND EXPOSURE

Only a small fraction of the organisms that live ever become fossilized, and almost all of those that do have hard parts such as bones or shells. The many with only soft parts are not preserved, although now and again we find an imprint of one of their bodies. Bony and shelled creatures therefore dominate the discovered fossil record. We find only a fraction, and an unknown fraction at that, of the complete record.

A different problem arises from the way in which rocks are exposed at the earth's surface. As the Grand Canyon shows so beautifully, sedimentary rock formations generally are horizontal or not far from it. They may extend in area for hundreds or thousands of miles. But where are such rocks exposed? Not along a horizontal surface. With few exceptions, there they are covered either with soil or by other rock layers. To observe bedrock, we usually have to find a spot where some human or natural agent, like the Colorado River, has made a vertical cut down through the rocks, exposing a cross section. Although rock formations extend for vast horizontal distances, they can be seen and sampled only here and there, in road cuts, quarries, river banks, sea cliffs, and so forth, and therefore we have access to but a tiny fraction of their true volumetric extent. The lack of rock exposure causes us to find fewer organisms than actually lived and therefore to underestimate the true ranges of species.

BIOTURBATION AND REWORKING

As discussed in an earlier chapter, many marine animals—clams, for example—burrow downward into the sediment beneath, dragging down younger material from the surface and bringing older material back up, an effect known as bioturbation. When these disturbed sediments eventually harden into rock, the fossils that they contain, as well as any iridium and tektite layers, are stretched out over a broader range than the one in which they were deposited. Studies of bioturbation in modern sediments have shown that material can be moved up and down by many centimeters, equivalent to tens of thousands of years.

Reworking is a similar, but mechanical, process in which a layer of sediment, and the biota it contains, is deposited in the sea or in a streambed but, before it is hardened into rock, is stirred by waves or currents, and redeposited. As with bioturbation, such disturbance before sediments are consolidated mixes them up and causes some of the temporal information to be lost.

Both bioturbation and reworking raise fossils from the dead, like zombies, and redeposit them higher in the section than they deserve to be, in younger rocks than those in which the organisms actually lived. This makes it appear that the organisms lived longer than they did and thus causes a sudden extinction to appear gradual, or not to have happened at all. For example, Tertiary sedimentary rocks sometimes contain foraminifera that some specialists believe lived only in the Cretaceous; to be present in Tertiary sediments, therefore, the forams must have been reworked. Others believe that instead these forams survived the K–T event and lived on, into the Tertiary, in which case they were not killed off by an impacting meteorite and the Alvarez theory is undercut. (In Chapter 9 we see how paleontologists go about trying to solve this particular puzzle.)

Channel Cutting and Deposition

Rivers move back and forth across their floodplains, eroding here and depositing there. They cut channels down into the rocks beneath them at one time and later deposit fresh river sediment into those channels. Rivers can dislodge fossils from the rocks along their beds and banks and deposit them in their water-cut channels. This effect, like bioturbation, juxtaposes material of different ages.

As long as the channel-deposited rocks can be distinguished from those the channel is cut into, no one is led astray, but if, say, both are sandstones, it may not be easy to tell them apart. When we do not recognize the channel deposits for what they are, younger fossils carried downward into older rock appear to belong there and to have originated before they actually did. In the famous dinosaur beds of Montana, for example, fossils of mammals that were to be important in the Tertiary have been said to occur well down in Cretaceous rocks, suggesting that the replacement of dinosaurs by mammals began well before K–T time, which likely means that the dinosaurs were going extinct long before the boundary. But if the mammal fossils were washed off a Tertiary landscape and deposited into channels cut by Tertiary streams down into the Cretaceous rocks below, then the mammals and the channel deposits in which they are found are Tertiary, not Cretaceous, and the early replacement is an illusion. Here is another difficult puzzle for paleontologists and sedimentary rock specialists.

SAMPLING EFFECTS

At the first Snowbird conference in 1981, paleontologists Phil Signor and Jere Lipps presented what has proven to be one of the most important papers in modern paleontology.[11] Like some others that have had such a result, their paper was short and simple. The authors showed that sampling can have two separate but related effects on paleontological evidence, both of which make it harder to draw firm conclusions.

The first is illustrated by Figure 18, adapted from their paper, which shows how the diversity of ammonites, the beautifully coiled and chambered marine fossils that grace natural history museums, waxed and waned during the Mesozoic era, which includes the Triassic, Jurassic, and Cretaceous periods; and how the extent of sedimentary rocks deposited during the Mesozoic also varied (the Mesozoic lasted from 250 million to 65 million years ago). Though the match is not perfect, the chart shows that the more rock exposed, the more diverse the ammonites appear to be and the less rock exposed, the less diverse. For example, ammonite diversity appears to have declined from the middle to the end of the Cretaceous. But so does the amount of rock deposited. Thus the apparent decline in diversity can be explained entirely by the decreasing amount of rock available to be sampled. Ammonites might have been thriving when, by coincidence, the amount of rock being deposited and preserving

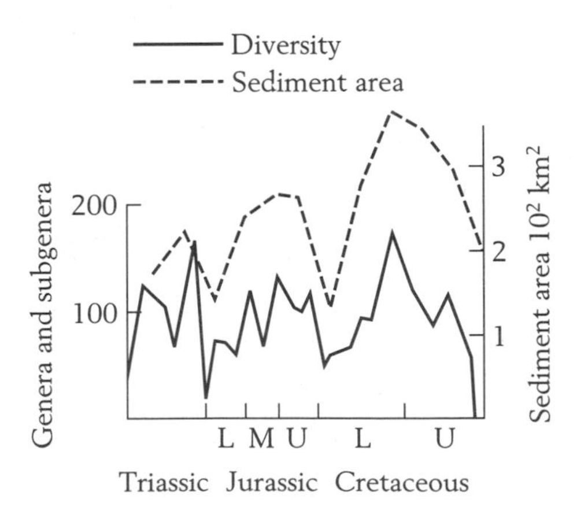

FIGURE 18 The diversity of ammonite genera and the area of sedimentary rock across the Mesozoic era. Note that the two roughly correspond. [After Signor and Lipps.[12]]

their remains was shrinking, causing us to conclude incorrectly that their true diversity had fallen. Of course, a direct link between the two might have existed: Whatever caused fewer deposits might also have caused ammonite diversity to decline, but that we cannot know. Signor and Lipps concluded that "diversity data cannot be taken at face value": The availability of rocks to be sampled can control the apparent abundance of fossils.[13]

This first effect has relevance to the diversity of dinosaurs. Several scientists have noted that there are fewer dinosaur species found in the last stage of the Cretaceous, called the Maastrichtian, than in the immediately older stage. (Each section of the geologic column is named either for a place where it was first recognized, or where it is thought to be particularly well exposed. In this case, the "type locality" is the Dutch town of Maastricht near the Belgian/German border.) This decline in species collected suggests that the dinosaurs were already on the wane by the middle and late Cretaceous, leaving nothing for meteorite impact but a possible coup de grace. But dinosaur specialist Dale Russell pointed out that since the Maastrichtian lasted for only about half as long as the Campanian, we would naturally expect it to produce only about half as many species.[14] This conclusion has been disputed, but the ammonite and dinosaur examples remind us that apparent changes in diversity may simply be artifacts of differing sample sizes. To the extent they are, we underestimate the true range of species and conclude they went extinct before they actually did.

In order to understand the second of the two effects pointed out by Signor and Lipps, imagine that you have to approximate the Canada–United States boundary using one of two methods: (1) by locating the houses of the northernmost residing United States citizens, or (2) by locating the houses of the northernmost residing members of Congress.[15] Obviously, using the abodes of citizens would give the more accurate result. Using the homes of the more rare congressional representatives would cause you to place the boundary further south than it really is. In the same way, the more rare a fossil species, the less likely we are to find its true geologic level of extinction. This gives rise to the "Signor-Lipps effect," a concept with which every paleontologist studying changes in diversity over time henceforth must wrestle.

Figure 19 represents a hypothetical cross section down through a formation that contains three fossil species, each marked by a different geometric symbol. The diagram assumes that each species became extinct at the same time, represented by the top of the draw-

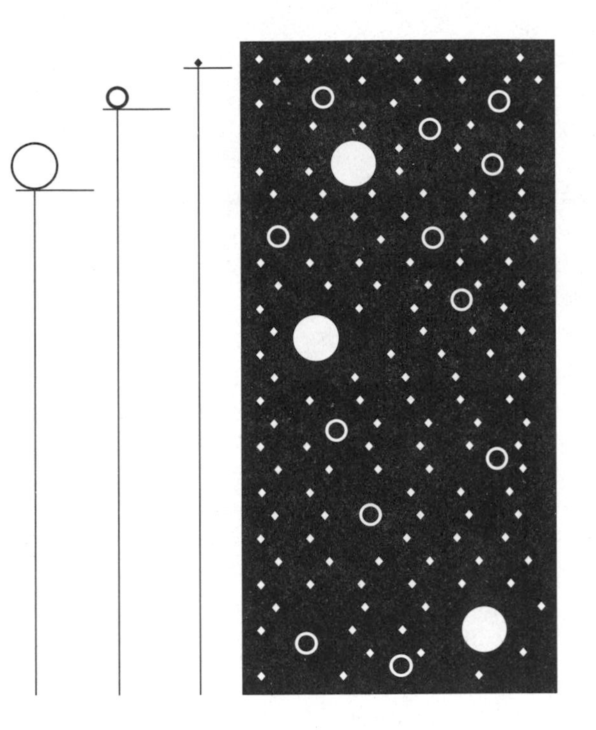

FIGURE 19 The Signor-Lipps effect. [After Michael Williams.[16]]

ing. Suppose that we sample every portion of this imaginary section of rock, missing nothing. What do we conclude? That the rarest fossil, shown by the large white circles, went extinct first, below its actual extinction level as represented by the top of the drawing. Because this species is rare, our chances of finding it anywhere, much less at its true level of extinction, are small. The next rarest species, shown by the small filled circles, appears to have become extinct a little higher up. The most common, marked by the small diamonds, is found right up to the "true" extinction boundary. Thus the "diversity" of species in this formation over time—the number preserved at each level in the rock—appears to have steadily decreased upward toward the extinction boundary, where all three actually disappeared. We conclude that extinction was gradual—no sudden disappearances here—but we are dead wrong: All three species lived right up to the boundary.

Now, factor in our inability to ever collect more than a fraction of the fossils present in a rock formation, by imagining that our sampling catches 10 percent of each species present. Mentally strike out, at random, nine of ten of the symbols representing the three fossil types and see what conclusion you would draw. You would miss all specimens of the rarest species (fewer than nine are present to start with) and conclude that it had become extinct even before this geologic section formed. The apparent level of disappearance of the other two species would move down in the section, causing you to place each of their extinctions at a level even further below their actual occurrence.

Thus the rarer a species and the less perfect the sampling, the earlier and more gradual its extinction appears. We see, not reality, but the false, gradual extinction of the Signor-Lipps effect. If each of the three symbols stands for a dinosaur species and the time period represented is the late Cretaceous, we would conclude that dinosaur diversity gradually declined and therefore that they were already doomed—no meteorite impact is required. But we would be wrong, victims of the Signor-Lipps effect.

This has been a thought experiment. What about a real one? A clever geologist named Keith Meldahl went, not back in time, but to a modern tidal flat in Mexico, where the muds are full of shelled marine species.[17] He imagined that an extinction suddenly occurred the day he visited the tidal flat, and that it was then preserved and sampled by some paleontologist far in the future. Meldahl drilled eight cores into the muds to a depth of about 70 cm and studied the extracted sediment centimeter by centimeter, making a careful record of each species and the highest point at which it was found. (Remember that all the species are alive today.) He located 45 different species in all; their positions in the cores are shown in Figure 20.

Even though this imaginary "extinction" was perfectly abrupt—far more so than any real extinction including that produced by impact—Meldahl actually observed the false, gradual pattern predicted by the Signor-Lipps effect. Of the 45 species that were present at the tidal flat, 35 appeared to go extinct below the surface, and this happened even though the cores were crammed full of "fossils-in-the-making," averaging almost 40 percent shell material by weight. In other words, the Signor-Lipps effect distorted the record even for relatively common species. Had the effect been forgotten with time, the paleontologist of the future, unaware, would naturally conclude that three out of four species had gone extinct gradually.

Sampling problems can never be entirely eliminated, but they lessen as more samples are collected, which is exactly what paleontologists have been doing since the Signor-Lipps effect was described. But no matter how exhaustive and exhausting the collecting, the inexorable mathematics of sampling means that some effect will always remain. "Gradual extinction patterns prior to a mass extinc-

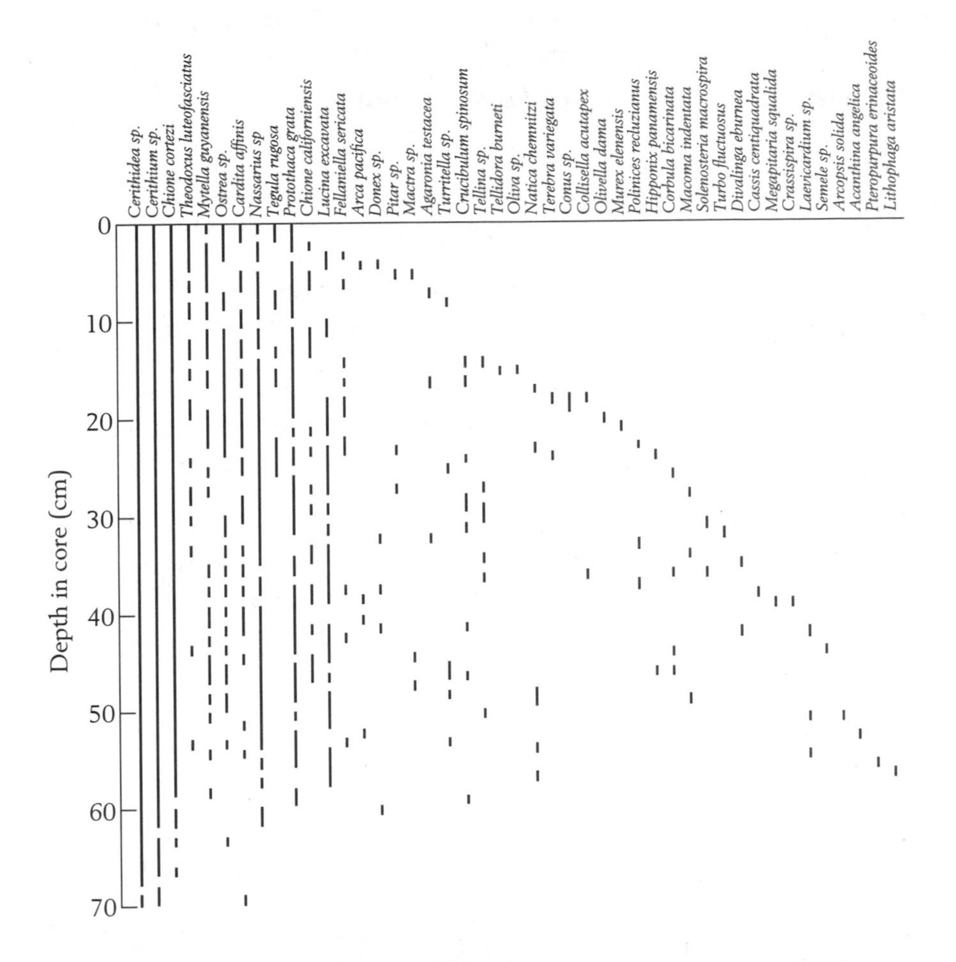

FIGURE 20 The Signor-Lipps effect in operation at a modern tidal flat. Depending on their rarity, species appear to become extinct at different depths in the core, equivalent to different times in the past. Yet each species is alive today. [After Meldahl.[18]]

tion do not necessarily eliminate catastrophic extinction hypotheses," Signor and Lipps concluded at Snowbird I. "The recorded ranges of fossils . . . may be inadequate to test either gradual or catastrophic hypotheses."

INTERPRETING THE FOSSIL RECORD

Remember that to test the Alvarez theory, we want to know whether fossil species were already on the way out in advance of the Chicxulub impact, and whether those that became extinct did so

TABLE 3

Problems of Interpreting the Fossil Record

Problem	Effect
Gaps	Gradual extinction appears sudden.
Migration	Range shortened; extinction seems to have occurred earlier than it did.
Poor preservation and exposure; dissolution	Range shortened; species appear more rare than they were; extinction seems to have occurred earlier than it did.
Bioturbation and reworking	Ranges extended upward and downward; sharp layering is smeared out. Reworked and survivor species confused. Sudden extinction appears gradual.
Channel deposition	Species appear too early; those that did not live together may be found together.
Reduced sample size	Sudden extinction appears gradual.
Signor-Lipps effect	Sudden extinction appears gradual.

exactly at the K–T boundary. To perform these tests, geologists needed to be able to pinpoint the exact level in a bed of rock at which a species became extinct. This brief review has shown what a formidable challenge such a requirement presents (summarized in Table 3). The fossil record begins in imperfection and is then altered by sedimentary and biologic processes; what was gradual may appear falsely sudden, what was sudden may appear falsely gradual; the highest fossil is never found.

In 1980, when the Alvarez theory broke upon the scientific world, geologists had already been aware of these problems (except for the Signor-Lipps effect) for a long time. Since they had not been focusing on extinction, however, they had not been attempting to determine exactly where a fossil species disappeared. Their initial response was to rely on the data already collected, and their prior interpretations of it, but those soon proved inadequate to provide a proper test. Paleontologists returned to the field to collect the much larger sample fossil databases that were required to probe the Alvarez theory. Gradually, enough data became available to allow researchers to test in detail its predictions for mass extinction.

CHAPTER 9

A WHIMPER OR A BANG?

We really hadn't been looking at the record in enough detail to pick this extinction up, and we weren't disposed to look at it as a catastrophe.[1]
Leo Hickey

It is easy to say that 70 percent of all living species became extinct at the K–T boundary, but what does it really mean? Who, and how many, died? Remember that with the exception of species, the taxonomic groupings established by biologists and paleontologists have no inherent meaning—they are just one of many ways of organizing flora and fauna. Phyla, families, groups, and genera do not die; they are artificial constructs of the human mind. Only living individuals die. How many have to do so before too few breeding pairs (if that is the way they do it) are left to allow a species to survive? Although we cannot be sure of the answer, the family hamster provides a familiar (too familiar?) example. David Carlisle has stated that as far as we know, "every golden hamster now alive is descended from a single pregnant female trapped in Syria early in this century."[2] In the case of the ubiquitous hamster, survival did not even require a pair! Try to imagine, if you are willing, what it would take to exterminate *Homo sapiens*. Only a few couples surviving in caves or in remote regions near the poles might carry us through, to begin again, as in Walter Miller's classic, *A Canticle for Leibowitz*.[3] To truly eliminate our species, would not nearly every human being on earth have to die?

This repugnant thought experiment and the hamster example may lend some credibility to Carlisle's claim that 99.99 percent of all individuals of every species alive before the K–T boundary—*including individuals of the species that survived*—died in the K–T event, leaving a minimum breeding population of no more than 0.01 percent. Carlisle does not say how he arrived at this figure, but his claim does provoke us into contemplating the enormity of the task of extinguishing entire genera—an almost unimaginable number of individuals must die. Although an event that kills such a high percentage of all living creatures is indeed nearly impossible for us to imagine, one would not want to be around to learn whether an explosion with the energy of 7 billion bombs the size of the one dropped on Hiroshima would do the awful job. Carl Sagan and the other modelers of nuclear winter feared that a set of explosions totaling only a fraction of the energy released by the Chicxulub impactor might cause the extinction of the entire human race.

Surely a dying massive enough to eliminate 70 percent of all species would leave in the fossil record clear evidence of its destructiveness. Whether it has—whether the fossil evidence corroborates or falsifies the Alvarez theory—hinges on two critical predictions. The first searches for evidence below the boundary; the second searches above it.

THE EFFECTS OF IMPACT ON LIFE

PREDICTION 1: Prior to the K–T boundary, most species were not already going extinct for some other reason. Their extinction was sudden and right at the boundary.

We can summarize this prediction by saying that the K–T impact will not have been "anticipated." That is, most species that did not survive the boundary were not already going extinct for some other reason; instead, their extinction was delayed until the impact and was caused by it. Critics of the theory argue that just the opposite is true. They say that the dinosaurs, for example, were already on the way out, and being replaced by mammals, well before the end of the Cretaceous. In their interpretation, impact may have happened, but if so it only finished off a few doddering stragglers. Thus meteorite impact has no appreciable effect on life and can be ignored as a factor in evolution. To test this first prediction, geologists need to find and to trace fossils of key species from levels well down in the Cretaceous right up to the boundary.

Of course, as paleontologists have known for over a century, some species did become extinct late in the Cretaceous, but before the boundary, as happens in any geologic period. Conversely, some made their first appearance then. The question is whether this was the case with the most prominent fossil families that we know did not survive the boundary, including the four that are the focus of this chapter: the ammonites, plants, forams, and dinosaurs. Were any of them already well on the way out long before the K–T meteorite made its appearance?

PREDICTION 2: Except where reworking has occurred, species that became extinct at the K–T boundary will not be found above the iridium horizon.

The Alvarez theory maintains that the primary lethal effects of the impact were immediate and that the secondary ones lasted for at most a few hundred or a few thousand years. Since on a geologic time scale these are instantaneous, the time of impact as located by the iridium horizon and the time of the mass extinction are the same. In effect, this prediction holds that the mass extinction, the iridium horizon, the clay layer, and the K–T boundary all are synchronous. Fossils of the major groups that became extinct at the K–T boundary, such as the ammonites and dinosaurs, or that experienced a major species turnover then, such as the foraminifera and plants, will not be found above the iridium level (unless they were brought there by bioturbation or reworking).

But suppose that a few species thought to have gone extinct at the boundary were to turn up above the iridium layer, in confirmed Tertiary rocks: Would that falsify this prediction? In fact, dinosaur fossils have been claimed from Tertiary rocks in China and Montana, though the claim naturally depends on exactly where scientists placed the K–T boundary in each locale and on the assumption that the fossils there are in their original geological sites. We know from their absence in the subsequent fossil record that dinosaur survival into the Tertiary would at best have been rare and temporary. The finding that a few species made it through, only to become extinct a short while into the Tertiary, would have no bearing on whether impact caused the mass extinction and would be nearly immaterial to broad earth history. If, however, the main extinction horizon for taxa (a catchall name for an unspecified taxonomic group: species, genus, family, order, etc.) that have always been regarded as having gone extinct at the K–T boundary—say the ammonites and dinosaurs and most plants and forams—were actually found to lie above the iridium, the case for impact-induced, large-scale extinction would be weakened. If all were consistently found above the iridium, this prediction would have failed and this half of the theory would be undone. We would know that a giant impact occurred at the end of the

Cretaceous and that something else caused the mass extinction; we would be back at square one.

OF AMMONITES, PLANTS, AND FORAMS

It is hard to think of four more diverse groups of organisms than the ammonites, plants, and forams plus the dinosaurs. The ammonites and forams both made their lives in the sea, but in completely different ways; both are fundamentally different from the plants. These four have now been studied sufficiently to allow us to use them to test the extinction half of the Alvarez theory.

AMMONITES

Ammonites were mollusks, like the squids and octopuses, that ranged from about a centimeter to a meter in size. Their intricately coiled and chambered shells provide some of our most beautiful fossils. The ammonites first appeared in the mid-Paleozoic, survived even the deadly end-Permian extinction, when 96 percent of all species died, and expired at the end of the Cretaceous, 330 million years after their arrival. Geologists, including Darwin (whose contributions to that field earn him the title geologist) for over a century had noted that the ammonite extinction marked the K–T boundary. Yet when modern geologists took a closer look, they were not so sure. Peter Ward of the University of Washington, a leading expert on ammonites, tells of being invited to Berkeley in 1981, not long after the Alvarez theory had appeared, to give a talk to assembled Alvarezes and Berkeley paleontologists.[4] He told them of a theoretical study he had just completed, which showed that the ammonites should have gone extinct suddenly. This was music to the ears of the Alvarezes, who lost no time in inviting Ward to dinner. According to Ward, William Clemens, a Berkeley paleontologist whose office was down the hall from Walter's and who was to become a bitter opponent of Luis's, left immediately after the talk without comment.[5] A year later, after having done extensive ammonite collecting, Ward returned to Berkeley to give another talk, this time announcing that his new findings showed that the ammonites had apparently gone extinct well below the boundary, thus contradicting prediction 1 and undercutting the Alvarez theory. This time Luis Alvarez stayed away, while Clemens invited Ward to dine.

One of the best locations for finding ammonites near the K–T boundary is in the Bay of Biscay off Spain, at its border with France, where the relevant geologic section is well developed and well exposed in huge, wave-cut sea cliffs. Ward's favorite outcrops were near the town of Zumaya, where a strong iridium anomaly was found in the same kind of thin clay layer as at Gubbio. In 1983, not long after the appearance of the Alvarez theory, Ward reviewed the life and death of these fascinating creatures in an article in *Scientific American*.[6] Because the highest (youngest) ammonite he could find occurred some 10 m below the K–T boundary—equivalent to tens of thousands of years before the impact—Ward concluded that the Alvarezes were wrong: "The fossil record suggests . . . that the extinction of the ammonites was a consequence not of this catastrophe but of sweeping changes in the late Cretaceous marine ecosystem . . . studies . . . at Zumaya suggest they became extinct long before the proposed impact of the meteoritic body."[7] At the end of his article, however, Ward added a crucial caveat: "This evidence is negative and could be overturned by the finding of a single new ammonite specimen."[8]

Ward's reference to negative evidence was meant to emphasize that the absence of evidence is not evidence of absence. Failure to find a fossil species at a given horizon near its upper limit does not prove that the organism had already gone extinct at that level, as Ward understood—maybe a more diligent search would turn it up. The only way to make progress against the Signor-Lipps effect is to return and search again. Ward returned to Zumaya to do exactly that. "Finally, on a rainy day," he writes, "I found a fragment of an ammonite within inches of the clay layer marking the boundary."[9]

Thus encouraged (and perhaps influenced by the appearance on the Zumaya beach first of armed Spanish soldiers and later of disgruntled Basques, each asking what he was doing to their rocks), Ward began to enlarge his collecting areas to include other sites along the Bay of Biscay where the K–T boundary is exposed. Without the impetus of the Alvarez theory, Ward would not have gone to this extra trouble. What would have been the incentive to spend more of his life on these same cliff faces? His redoubled effort paid off. Within the first hour of collecting near the French town of Hendaye, just around the corner from Spain, he found abundant ammonites in the last meter of Cretaceous rock. In papers given at the second Snowbird conference in 1988, Ward reported that "collecting...east and west of the Zumaya section ultimately showed that ammonites are relatively common in the last meter of the Cretaceous sediment."[10] He thought that their scarcity at Zumaya had stemmed "from some aspect of ammonite ecology, rather than collection failure or preservation effects."[11] For now vanished reasons, the ammonites had migrated away from Zumaya, but had not gone far. Only a short distance away, at Hendaye, they showed up in abundance. In 1994, Ward summed up: "My decade-long study . . . in Spain and France ultimately showed that the ammonites had remained abundant and diverse right up until the end of the era; the last ammonites . . . were recovered just beneath K–T boundary clay layers."[12]

Ward's latest approach to the problem of ammonite extinction is to apply statistics. Considering the difficulties inherent in trying to locate the true extinction horizon of a species, using statistics not only makes sense, it is essential. At best, for a given horizon in time, scientists have access to only a tiny fraction of the geologic record. Combine that with the Signor-Lipps effect, and you can see that the chances are vanishingly small that the last surviving individual of any fossil species—say, the last ammonite to have lived and be fossilized—will ever be found. The time of the true, last survivor will always lie above the horizon of the highest specimen recovered. One way around this difficulty is to collect large numbers of samples and then to apply statistical techniques, as Ward and Charles Marshall of UCLA have done for the ammonites at Zumaya.[13] This allowed them to define a range over which a given ammonite species most probably became extinct. They found that a few ammonite species disappear all the time in a kind of background extinction, and that, prior to the K–T boundary, a drop in sea level apparently had killed off a few others, though this happened gradually. They concluded that 50 percent of the Zumaya ammonite species had undergone a sudden extinction right at the boundary.

What lessons can we learn from the story of the Biscay ammonites? Most important, and contrary to the initial impression of Ward and others, half the ammonite species lived right up to the K–T boundary, when they suffered an extinction that was both sudden and catastrophic, corroborating prediction 1. Ammonites are not found above the iridium, corroborating prediction 2. Although their numbers waxed and waned throughout the Cretaceous, as they had done for the preceding 330 million years, as nearly as we can tell given the vicissitudes of the fossil record, the final demise of the ammonites is fully consistent with the Alvarez theory.

PLANTS

The accumulation of knowledge has meant that scientists today must practice in finer and finer subspecializations: One is not a paleontologist, one is an invertebrate paleontologist specializing in ammonites; or one is a palynologist, an expert not just in fossil plants, but in fossil pollen spores. One of the most beneficial by-products of the Alvarez theory is the way in which it has brought together scientists from an unprecedented variety of disciplines. Pollen specialists, for example, have found themselves for the first time in the same room with dinosaur experts, chemists, physicists, and astronomers, all discussing supernovae, precious metals, impact explosions, and mass extinctions. Advances have been made that would have been impossible had only one group been involved. In this sense, few theories in the history of science have been as fertile as the Alvarez theory.

Two vertebrate paleontologists, William Clemens and David Archibald, along with plant specialist Leo Hickey, were among the first to speak up in opposition to the Alvarez theory, though in much less detail than did Officer and Drake. Only months after the original Alvarez paper appeared in *Science*,[13] Clemens, Archibald, and Hickey published "Out With a Whimper Not a Bang."[14] This 1981 paper and its title have become metaphors for the initial reaction of paleontologists to the proposal that meteorite impact caused the K–T mass extinction. They closed with this paraphrase of T. S. Eliot's lines:

This is the way Cretaceous life ended,
Not abruptly but extended.[15]

Although plants have received far less notice than the more fascinatingly popular dinosaurs, paleontologists have known for a long time that many plant species also failed to survive the K–T boundary. The lack of attention is unfortunate, for fossil plants can tell us a great deal about mass extinctions. First, as the base of most food chains, plants determine much of what happens in the entire realm of biology. Second, because they are so different from animals and are sensitive environmental indicators, fossil plants reveal a lot about the nature of extinction events. Third, pollen and leaf fossils can be present in large numbers, reducing sampling errors and allowing the statistical techniques that add confidence to conclusions about extinction rates and timing.[16]

"On the whole the pattern of change in land plants and the increasingly cooler affinities of the latest Cretaceous to early Paleocene [earliest Tertiary] palynofloras [pollens] are compatible with a gradualist scenario of extinction possibly related to climatic cooling," Clemens, Archibald, and Hickey concluded.[17] That same year, Hickey wrote that the evidence from land flora, together with the difference in the time of extinction of plants and dinosaurs, "contradict hypotheses that a catastrophe caused terrestrial extinctions."[18] He based this conclusion on some 1,000 leaf fossils that he and his colleagues had studied, which at the time seemed a large sample indeed.

The first piece of evidence to suggest that the conclusion might be wrong, or at least not universally applicable, came in the 1981 paper by Carl Orth and colleagues in which the first iridium spike was reported in nonmarine rocks from the Raton Basin, proving that the iridium had not been concentrated from seawater.[19] They also found that right at the level of the K–T boundary and the iridium spike, the pollen of angiosperms—the flowering plants—nearly disappeared, while that of the ferns rose dramatically. This "fern spike" subsequently turned up at several other K–T localities and in various rock types. Botanists know from studies of modern catastrophes—from the eruptions of El Chichón, Krakatoa, and Mount St. Helens, for example—that ferns are opportunistic plants that move in quickly to colonize a devastated area. Flowering plants later replace them, as happened in the early Tertiary. This scenario suggests that for the flowering plants, the Cretaceous ended not with a whimper but with a bang, quite abruptly.

At the Snowbird II conference in 1988, Hickey and Kirk Johnson reported the results of a new study of nearly 25,000 specimens of mainly leaf fossils from more than 200 localities in the Rocky Mountains and the Great Plains.[20] The 25-fold increase in the number of specimens collected over the original Hickey study reflects the impact of the Alvarez theory. Hickey and Johnson found that 79 percent of the Cretaceous plants had gone extinct at the K–T boundary, at the same point at which the fossil pollen changes and the iridium spike appear. This new and statistically more sound evidence caused Hickey, like Peter Ward, to change his mind and conclude that "The terrestrial plant record [is] compatible with the hypothesis of a biotic crisis caused by extraterrestrial impact."[21]

Speaking to *Science* reporter Richard Kerr in 1991, Hickey was bluntly honest: "I became a believer. This evidence is incontrovertible; there was a catastrophe. I think maybe [the anticatastrophism] mind set persisted a little too long."[22] Like most paleontologists, Kirk Johnson was initially "skeptical of this outlandish theory that attributed the demise of our beloved dinosaurs to some science fiction asteroid."[23] His own studies of leaf fossils from the dinosaur beds of Montana made a believer out of him. Archibald also appears to have been converted, at least on the plant evidence, writing in his 1996 book, "Of all the data from the terrestrial realm, the record of plants in the Western Interior seems to me to present the strongest case that extinction was rapid, not gradual, for the species so affected."[24]

As Johnson has continued to sample the fossil plant record in Montana and North Dakota, he has found more and more new plant species in the uppermost Cretaceous, but next to no new Tertiary ones. This means that more Cretaceous species died out than he had measured earlier. Johnson now estimates the percentage extinction at close to 90 percent.[25] Thus the fossil plant evidence thoroughly corroborates predictions 1 and 2: As many as 90 percent of Cretaceous plant species disappeared suddenly, right at the K–T boundary; none of them are found above the iridium level.

FORAMINIFERA

Planktonic foraminifera (Figure 21), nicknamed forams, are one-celled, amoebae-like protozoa that float at various depths in the oceans, eating the still smaller photosynthetic algae and secreting calcareous shells that survive the foram's demise. They evolve rapidly

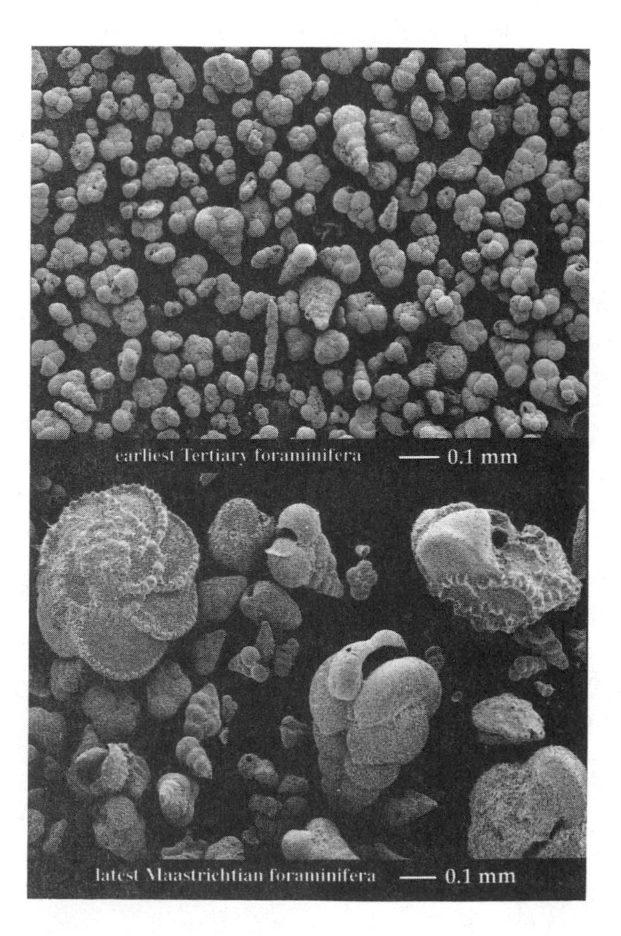

FIGURE 21 Foraminifera. [Photo courtesy of Brian Huber and National Museum of Natural History.]

and spread out over wide areas, making them handy for indexing rock formations. Their sensitivity to temperature and salinity also make them useful in interpreting ancient environments. Foraminifera occur in the hundreds of thousands, even in a single hand-sized specimen.

At the time the Alvarez theory appeared, the paleontologists who had been studying fossil plants believed they had largely survived the K–T boundary; those who had been studying ammonites believed they had gone extinct well before it. A return to the field for more collecting showed that both interpretations were wrong: The ammonites and the plants each suffered a massive extinction right at the K–T boundary. In contrast, the intensive study of the forams in the years following the appearance of the Alvarez theory led some paleontologists to just the opposite conclusion: What appeared to be a mass extinction was, they claimed, something else.

In 1980 few doubted that the foram extinction had been nearly complete and had been timed exactly to the K–T boundary; indeed, so many foram species disappeared that their level of departure almost defined the boundary. At Snowbird I, Hans Thierstein of the Scripps Institute of Oceanography showed that over 97 percent of foram species and 92 percent of the genera became extinct at the K–T boundary,[26] and Jan Smit reported that the K–T foram extinction was so thorough that only one species, *Guembelitria cretacea*, survived, with all the subsequent foram species having evolved from it.[27]

The puzzle for students of earth history is how creatures that made their living floating in the sea all could be killed at once. A clue comes from another group of forams—the benthic variety that live on the seafloor—which suffered a much lower rate of extinction at the K–T boundary. Some believe that the difference in survival rate stems from the dependence of the floating plankton on "primary productivity"—that is, they ate the even smaller plantlike phytoplankton and therefore would die if those organisms were not available. The benthic forams, on the other hand, lived down in the detritus of the seafloor where they could feed on the accumulated organic debris, which would have been abundant after so many other creatures, including their floating cousins, had died in the K–T extinction. Thus if the upper layers of the oceans became sufficiently poisoned to kill the phytoplankton, the floating forams would die out but the deeper benthic variety would live on.

The simple picture of nearly complete foram extinction right at the K–T boundary did not go long unchallenged. Gerta Keller, who emerged as Smit's leading opponent in the interpretation of K–T microfossils, studied forams at the K–T sections at El Kef, Tunisia, and along the Brazos River in Texas. At Snowbird II, Keller reported that "planktonic foraminifera show 30–45% of the species disappearing during the 300,000 to 400,000 years prior to the K–T boundary. . . . [They] show an extended K–T boundary extinction pattern beginning below and ending well above the boundary."[28] If confirmed, this would falsify both predictions 1 and 2 for the forams. But Smit, in contrast to Keller, did not find that any forams disappeared before the boundary.

Here we have an impasse. Two reputable scientists, each examining the fossils from the same section of rock, come to entirely different conclusions. What to do? The answer was to conduct a blind test in which samples are carefully collected under the supervision of a neutral party and then distributed to other experts, who identify the fossils without knowing from where in the section they come. New samples were collected at El Kef and distributed by Robert Ginsburg of the University of Miami to four independent specialists, not including either Keller or Smit.

Ginsburg was to present the results of the blind test on the El Kef samples at Snowbird III. Keller had departed the night before, and Ginsburg, returning from the excursion to Mimbral, fell down an escalator. He prevailed upon Fischer to present the results and flew home.[29] Richard Kerr reported that, after Snowbird III, "both sides claimed victory."[30] Keller argued that each of the four investigators had found that at least some fraction—ranging from 2 percent to 21 percent—of the Cretaceous forams had gone extinct before the boundary, which essentially confirmed that the extinctions were gradual. But Smit disagreed, telling Kerr that this was a typical Signor-Lipps effect. Smit then lumped together the results from all four investigators, but using only species that at least two of them had found. Each species that Keller said had disappeared before the boundary, Smit's technique located in the last sample immediately below it. Smit summed up: "Taken together, they found them all. This eliminates any evidence for preimpact extinction in the [open ocean] realm."[31]

Keller responded that if some of the investigators had been mistaken in their identifications and had lumped together species that looked similar but were not, then what was actually a series of gradual extinctions would appear to have been sudden. But as Kerr reports, Keller's own taxonomy came into question. Brian Huber of the U.S. National Museum of Natural History had studied and written about the forams in a deep-sea sediment core (from Ocean Drilling Program Site 738), on which Keller subsequently published.

"None of her taxonomy or quantitative studies can be reproduced," said Huber; "the gradual side of the debate doesn't hold water because of her inconsistencies."[32] Keller responded in a letter to *Science* in which she cited 13 errors or misstatements in Kerr's article.[33] For one: "It was I who could not confirm Brian Huber's . . . study rather than the reverse. . . . Huber's comments are therefore not likely to have been objective."[34] Kerr responded: "By combining the efforts of all four blind testers, Smit intensified the search until all of Keller's gradually disappearing species were found to persist up to but not beyond the impact."[35]

The results of the El Kef blind test were finally published in 1997.[36] Not surprisingly, Keller and Smit continued to disagree. Thus the notion that a blind test can resolve disputes of this kind seems not to be borne out in practice. Even when fossils are as abundant as the forams, uncertainty remains.

The key point in the dispute between Keller and Huber was the identification in the core from Ocean Drilling Program (ODP) Site 738 of specimens of a particular foram species, *Parvularugoglobigerina eugubina*. Though Keller said that *P. eugubina* was common in the core from ODP Site 738,[37] Huber could not find it.[38,39] To resolve the dispute, Huber asked Keller for permission to visit her laboratory so that she could point out *P. eugubina* to him in her samples. She agreed and Huber set off from his home base in Washington, D.C., to Keller's lab at Princeton, where he was joined by paleontologist Chengjie Liu of Rutgers. Even under Keller's supervision, however, they could not find *P. eugubina* in Keller's slides, and, according to Huber, she refused to show them the most critical samples.[40] Thus with regard to *P. eugubina,* it was Huber who could not confirm Keller's taxonomy, not the other way round, as she had claimed in her response to Kerr.

After Huber returned from "the worst scientific experience of his life," he learned that Keller and Norman MacLeod, who was also present at the Princeton meeting, had resorted to an unusual course of action.[41] Taking their disagreement with Huber outside the pages of journals, they went to the top, writing to none other than the secretary of the Smithsonian Institution, Robert McCormick Adams, to complain of Huber's behavior and to ask for the loan of certain specimens, requesting that the loan be handled by some other Smithsonian paleontologist than Huber.[42] Adams replied that he preferred to see such differences resolved "through the normal channels of scholarly discourse."[43]

MacLeod and Keller went on to co-edit *Cretaceous-Tertiary Mass Extinctions: Biotic and Environmental Changes,* which contains

20 individual scientific papers.[44] None of Keller's critics contributed an article; half of the chapters are authored or co-authored by prominent critics of the Alvarez theory—Keller, MacLeod, Stinnesbeck. In the book, MacLeod and Keller sum up the foram evidence by repeating their claim that two-thirds of the species went extinct before the K–T boundary and the other one-third survived it. This hearkens back to the original argument of Officer and Drake that the change from the Cretaceous to the Tertiary was not instantaneous, but took place over an interval of time.

Most of the "Cretaceous" foram fossils that Huber studied from ODP Site 738 he found to persist above the K–T boundary. Did these species survive into the Tertiary or were they reworked? That is the question. If large numbers made it through the K–T event to live on in the Tertiary, they were not killed by a meteorite impact at K–T time but died later from other causes, falsifying prediction 2 for the forams. Thus the key question is whether the "Cretaceous" foram species found above the K–T boundary had already died out and were reworked into the Tertiary, or whether they actually survived the boundary event to die thousands of years later. Keller and MacLeod have addressed this question in a series of papers based on studies of K–T sections from around the world, testing for survivorship using techniques from paleontology, biogeography, and geochemistry. ODP Site 738 was among their most thoroughly studied cases. They concluded that there is no causal link between mass extinction event and direct effects of K–T boundary impact.[45]

As a youngster growing up on a farm in northern Ohio, Brian Huber could never have imagined himself on a research vessel in a spot so remote as to be called Desolation Island. Better known as Kerguelen, this tiny dot just off Antarctica, deep in the southern Indian Ocean, was long an important stop for whalers and seal hunters. (For a superb fictional account, read Patrick O'Brian's *Desolation Island*.[46]) Huber was there as a paleontologist on Leg 119 of the Ocean Drilling Program, which sailed from Mauritius in December 1987. During Leg 119, hole 738 was drilled in the seafloor off Kerguelen, giving Huber an enduring interest in this site. When the voyage began, like most paleontologists, he was dubious about the Alvarez theory. As he tells it, however, when the section of the core that traversed the K–T boundary was drawn to the surface and laid out on the deck, his doubts vanished. There, as at Gubbio, was the dramatic color contrast between the white, foram-rich sediment below and the fossil-poor, reddish section above. You could lay a knife blade right on the boundary, exactly where an iridium spike of 18,000 ppt, one of the highest ever measured, was later found. The boundary clay displays fine laminations that would have been destroyed by bioturbation, allowing that effect to be ruled out in this case.

Over the years, Huber and his colleagues have conducted detailed studies of the core from ODP Site 738.[47] They have found that specimens of some foram species that are known to have gone extinct at the K–T boundary, and of specimens of inoceramids, a group of clamlike creatures that became extinct well below the boundary, occur in the core far above the boundary, where they could have gotten only by reworking. Thus, in this core, contrary to the claims of MacLeod and Keller, reworking is prominent. This point is further supported by studies of strontium and carbon isotope ratios within the core. Huber concludes that "the high occurrences of Cretaceous species in lower Paleocene [lowest Tertiary] sediments are likely the result of extensive reworking."[48] He believes that only two of the Cretaceous species in the core from ODP Site 738 survived the K–T boundary and that the rest are reworked. But if MacLeod and Keller's methods for detecting reworking can err in this most studied of deep-sea cores, how do we know they are not in error in the other cores and geologic sections that they have studied and pronounced free of reworking? The two of them appear to be almost alone among micropaleontologists in denying that the foram evidence corroborates the impact-extinction link. But let us remind ourselves that we are not trying to prove that link, we are asking whether the foram evidence falsifies prediction 2, that species that became extinct at the K–T boundary will not be found above the iridium horizon, except where reworked. The majority of opinion is that the evidence does not.

What of prediction 1, that species will not have begun to go extinct prior to the boundary and will have disappeared suddenly? Keller, with MacLeod, continues the claim she made at Snowbird II and repeatedly since: The foram extinctions started as early as 300,000 years before the K–T boundary. But once again we find ourselves confronting the dreaded Signor-Lipps effect, which inexorably causes a sharp extinction to appear gradual, even at a modern tidal flat, where all the "extinctions" took place in an instant. How can it be ruled out in the case of the forams? Prediction 1 is not falsified by the foram evidence.

THE FOSSILS SPEAK

The history of understanding the ammonite and fossil plant extinctions at the K–T boundary shows how absent or negative evidence—

at first interpreted as falsifying the Alvarez theory—dissipated as more and more data were collected. Time and again, when paleontologists returned to the outcrop, they found what had eluded them before. In contrast, as more data were collected, the foram evidence seemed only to grow more complicated. But today, with a few noted exceptions, most specialists believe that the foram evidence is also compatible with an impact-induced extinction. I conclude that the evidence from all three groups—ammonites, plants, and forams—corroborates predictions 1 and 2: (1) Prior to the K–T boundary, most species were not already going extinct; rather their disappearance was sudden and right at the boundary. (2) Except where reworking has occurred, species of taxa that did not survive the K–T extinction are not found above the iridium horizon. Now it is time to turn to the dinosaurs, the creatures that got us interested in the K–T boundary in the first place.

Chapter 10

The Death of the Dinosaurs

... the impact theory of extinction? It's codswallop.[1]
William Clemens

In an article written in 1990, Michael Benton of the University of Bristol in England divided the history of dinosaur extinction studies into three phases.[2] From the time the existence of the terrible lizards was first acknowledged in the 1840s until around 1920, their extinction was a "nonquestion": The great, lumbering, pea-brained beasts had simply lost the survival race to the more nimble and intelligent mammals—our ancestors. During the "dilettante phase" from 1920 to 1970, interest in dinosaur extinction rose, and many theories were proposed, some of them downright silly, as the quotation from Glenn Jepsen in the Prologue makes clear. During this phase, dinosaur extinction appears to have been treated, sometimes by otherwise serious scientists in respectable journals, as little more than a parlor game. Perhaps this was a defensive mechanism: Unable to explain with any significant evidence the most notable of biologic and geologic mysteries, we masked our inability by trying to turn the whole matter into a joke.

The "professional phase" of dinosaur extinction studies began about 1970; by 1980, when the Alvarez theory appeared, most paleontologists had already made up their minds. At Snowbird I, the late Tom Schopf, yet another fine paleontologist from the University of Chicago, spoke for the majority: "A satisfactory explanation of the cause of the extinction of the dinosaurs has been known for some years. . . . Probably more than 99.99999% of all the species that have ever existed on Earth are now extinct. . . . The dinosaurs are among these. Extinction is the normal way of life. . . . As far as is currently known, it does not seem necessary to invoke an unusual event to account for the demise of the dinosaurs."[3] Writing in 1982, Archibald and Clemens used different words to make the same point: "At present, the admittedly limited, but growing, store of data indicates that the biotic changes that occurred before, at, and following the Cretaceous–Tertiary transition were cumulative and not the result of a single catastrophic event."[4]

ACRIMONY

In October 1982, only two years after the original paper appeared in *Science* and before most paleontologists had even begun to take it seriously, Luis Alvarez gave a long, detailed, and unusually personal talk at the National Academy of Sciences, the most prestigious invitational scientific society in the world, of which he was a member.[5] When his remarks appeared in print, they offended paleontologists, geologists, and others who preferred to see a certain level of polite discourse maintained in science. He had begun his talk with a preemptive declaration of victory: "That the asteroid hit, and that the impact triggered the extinction of much of the life in the sea—are no longer debatable points."[6]

Writing about a field trip to Hell Creek, Montana, source of *Tyrannosaurus rex* and the bedrock of dinosaur studies, Alvarez noted that "[The husband of one of his co-workers] tripped over a previously undiscovered *Triceratops* skull on one occasion. So we have not been a group of people each working in his own little compartment, but rather we have all thought deeply about all phases of the subject."[7] To suspicious geologists, this casual statement implied that he thought that their field was so easy that complete novices could not only stumble across new discoveries, they could solve persistent problems with thought that, however deep, could not, and need not, have gone on for long.

"A physicist can react instantaneously when you give him some evidence that destroys a theory that he had previously believed in. But that is not true in all branches of science, as I am finding out," Luis claimed.[8] Not every physicist reacted as he described, however. Less than a year later, astrophysicist Robert Jastrow, a professor at Dartmouth College, wrote: "So there we are. The asteroid theory was very attractive because it explained so much in a simple way, and many people will regret its passing. However, the evidence against it is very strong."[9] Thus as early as 1983, two physicists, with the utmost confidence, came to exactly the opposite conclusion about a matter of geology.

Where do physicists gain the self-assurance to make pronouncements in a field in which they have little or no training and experience? Walter Alvarez has observed that science is a hierarchy from the sophisticated and mathematical to the complex and nonmathematical.[10] The order runs, roughly: mathematics, physics, chemistry, astronomy, geology, paleontology, biology, psychology, and sociology. Fields high in the hierarchy use mathematics to explain the laws that they have derived. Though the discovery of these laws may require great feats of intellect, the laws themselves can often be simply stated ($E = mc^2$). Fields lower down deal with history, life, the human brain and behavior, which cannot be described mathematically or simply (with a few exceptions, such as plate tectonics). Scientists in these fields must handle (literally) dirty rocks and messy, squishy things like whole organisms. Physicists expect that someday they will be able to roll everything into one grand unified theory; scientists from geology "on down" would scarcely dream of such an aspiration. All this translates into a false sense of superiority on the part of those at or near the top of the hierarchy and gives them, seemingly without a moment of doubt, the nerve to make pronouncements about the fields below them. Ironically, in his article Jastrow acknowledged this hierarchy, but he failed to consider that it might apply not only to Luis Alvarez, but to himself.

Of course, as with other forms of prejudice, such attitudes are wrong, even dangerous. Mathematics is not "better" than psychology; it is merely different. Not every physicist could be a successful biologist. Just as we are finding that different kinds of intelligence exist, so each field probably attracts those most amenable to its special set of problems, techniques, and ways of thinking.

At the time of Luis's talk at the National Academy, plant paleontologist Leo Hickey was arguing that the plant extinction had been moderate and had occurred at a different time than the dinosaur extinction. After describing Hickey as a "very good friend" and a "close personal friend" of Walter Alvarez, Luis said that "Hickey has behaved quite differently with respect to the [Raton basin fern spike] . . . he ignored it."[11]

Alvarez directed most of his disdain at William Clemens, the vertebrate paleontologist and faculty colleague of Walter Alvarez, whom Luis also described in his talk as a friend. Their difference centered on the interval between the K–T boundary at Hell Creek, Montana, and the highest recovered dinosaur bone, which occurred at some distance below the boundary, producing a barren interval that became known as the "ghastly 3-m gap." Luis Alvarez describes in great detail how he used a variety of techniques to convince

Clemens that such a gap was only to be expected when rare creatures had suddenly gone extinct, yet Clemens stubbornly refused to accept the obvious. Luis's description of his attempt to persuade his "friend" goes on for four dense pages; its detail suggests that there must have been something more than friendship and science behind it—Alvarez seems to be trying to show Clemens up not only as wrong but as unreasonable, even unscientific: "I really cannot conceal my amazement that some paleontologists prefer to think that the dinosaurs, which had survived all sorts of severe environmental changes and flourished for 140 million years, would suddenly, and for no specified reason, disappear from the face of the earth . . . in a period measured in tens of thousands of years. I think that if I had spent most of my life studying these admirable and hardy creatures, I would have more respect for their tenacity and would argue that they could survive almost any trauma except the worst one that has ever been recorded on earth—the impact of the K–T asteroid."[12]

Shortly after he gave his talk, the review article by Archibald and Clemens in *American Scientist*[13] appeared, just in time for Alvarcz to incorporate a critique of it as an afterword to his written remarks. According to Alvarez, after they had ignored the iridium evidence and pooh-poohed impact, Archibald and Clemens offered only two alternatives as the cause of the K–T extinction: supernova explosion and the spillover of Arctic seawater (which would have lowered global temperatures). Alvarez said that he was "quite puzzled to see that in 1982, two knowledgeable paleontologists would show such a lack of appreciation for the scientific method as to offer as their only two alternative theories to that of the asteroid, a couple of outmoded theories. . . . Today, both of them are as dead as the phlogiston theory of chemistry."[14] To accuse a scientist not only of being wrong, but of being ignorant of the proper use of the scientific method, is the deadliest of scientific insults—tantamount to saying that the person in question is not truly a scientist. Such a remark would certainly strain any friendship.

Polls do not decide scientific matters, but since our interest is not only in what caused the K–T extinction, but in how scientists reacted to the Alvarez theory, it is useful to consider the results (normalized to equal 100 percent) of a poll taken in the summer of 1984 of over 600 paleontologists, geophysicists, and other geologists from six countries:

- 24 percent agreed that an extraterrestrial impact at the K–T boundary caused the mass extinction.
- 38 percent thought that a K–T impact had occurred but that other factors caused the mass extinction.

- 26 percent thought that no K–T impact had occurred.

- 12 percent believed that there had been no K–T mass extinction.[15]

As David Raup points out, over 60 percent of the polled scientists believed that an extraterrestrial impact ended the Cretaceous. Not bad considering the nearly 200-year influence of uniformitarianism.[16]

When the Society of Vertebrate Paleontologists held its annual meeting in Rapid City, South Dakota, in 1985, the controversy was in full bloom. According to reporter Malcolm Browne, writing in the *New York Times*, the assembled paleontologists claimed that the argument over impact had "so polarized scientific thought that publication of research reports has sometimes been blocked by personal bias."[17] (The discussion in this and the following three paragraphs is taken from Browne's article.) One said that "Scientific careers are at stake."[18] Some linked the cosmic winter that might result from meteorite impact with the nuclear winter that might result from World War III. Those who denied that cosmic winter could have occurred might also deny nuclear winter, thus branding themselves as pronuclear militarists.

Here among their own, not far from the best dinosaur collecting fields in the world, the vertebrate paleontologists let loose. The general thrust of the comments, though not the polite tone, was expressed by Robert Sloan of the University of Minnesota: "My own analysis of the fossil record suggests that the Cretaceous extinctions were gradual and that the catastrophic theory is wrong."[19]

William Clemens announced that he had discovered dinosaur fossils near Prudhoe Bay in Alaska, which, as it is today, was in the Arctic during the late Cretaceous and therefore subject to long periods of darkness. If dinosaurs could survive six months of Arctic winter and darkness, how could a few months of alleged cosmic winter kill them off? Said Clemens, "But survive they did, as we see in the fossil record."[20]

The most vicious attack came from Robert Bakker, formerly of the University of Colorado Museum, the originator of the theory that the dinosaurs had been warm-blooded and fast moving: "The arrogance of those people is simply unbelievable," he said of the proimpactors. "They know next to nothing about how real animals evolve, live, and become extinct. But despite their ignorance, the geochemists feel that all you have to do is crank up some fancy machine [presumably the iridium analyzer] and you've revolutionized science. The real reasons for the dinosaur extinctions have to do with temperature and sea level changes, the spread of diseases by migration and other complex events. But the catastrophe people don't seem to think such things matter. In effect, they're saying this: 'We high-tech people have all the answers, and you paleontologists are just primitive rock hounds.'"[21]

Luis Alvarez liked a fight and gave as good as he got. In a second article in the *New York Times,* Browne quoted from Luis Alvarez's just published autobiography: "I don't like to say bad things about paleontologists, but they're really not very good scientists. They're more like stamp collectors."[22] (Showing that they did agree on something, Jastrow had a couple of decades earlier compared geology to the collecting of butterflies and beetles.)

Alvarez was echoing the great British nuclear physicist and Nobelist, Ernest Rutherford, who divided science into physics and stamp collecting.[23] Rutherford's offensive statement may have stemmed from a burst of professional pride and can be excused as such. Alvarez's remark, on the other hand, like Bakker's, seemed much more personal and demeaning of an entire field of scholarship.

It got worse. Jastrow told Browne, "It is now clear that catastrophe of extraterrestrial origin had no discernible impact on the history of life as measured over a period of millions of years."[24] Alvarez retorted: "There isn't any debate. There's not a single member of the National Academy of Sciences who shares Jastrow's point of view. Jastrow, of course, has gotten into the defense of Star Wars, which for me personally indicates he's not a very good scientist. In my opinion, Star Wars doesn't stand a chance."[25]

Jastrow rejoined by pointing out that Alvarez had flown on the companion plane to the *Enola Gay* in the raid that destroyed Hiroshima and had been one of only five physicists willing to appear before the Atomic Energy Commission to denounce as a security risk Dr. J. Robert Oppenheimer, who had headed the Manhattan Project, on which he had been Alvarez's superior. In his autobiography, however, Luis Alvarez wrote that he had told the Oppenheimer inquiry panel that he had no doubt of Oppenheimer's loyalty to the country.[26,27] The Alvarez-Clemens debate continued in Browne's article, with Alvarez saying that "he considers Clemens inept at interpreting sedimentary rock strata and that his criticisms can be dismissed on grounds of general incompetence."[28]

Tragically, as this debate sank to ever lower depths, Luis Alvarez discovered that he had terminal cancer of the esophagus. He told an interviewer, "I can say these things about some of our opponents because this is my last hurrah, and I have to tell the truth. I don't want to hold these guys up to too much scorn. But they deserve some scorn, because they're publishing scientific nonsense."[29]

Luis Alvarez died on September 1, 1988, ending one of the most versatile, successful, and combative careers in modern science. At Snowbird II, held just six weeks later, one of the participants proposed two minutes of silence in his honor. Walter rose to say, "My father would have been mortified. He'd much rather have a good fight in his memory."[30]

THEORIES OF DINOSAUR EXTINCTION

If impact did not kill the dinosaurs, what did? Michael Benton, writing nearly three decades after Jepsen, found a total of about 65 seriously proposed ideas.[31] (He omits such recent suggestions as AIDS and terminal constipation.) Taking a very coarse cut, Benton's list can be aggregated as follows:

- Medical problems ranging from slipped discs to disease.
- "Evolutionary drift into senescent overspecialization." (Some of us may feel that we too are suffering from this malady—a sort of reptilian chronic-fatigue syndrome.)
- Competition with other animals, especially mammals.
- Floral changes: New plant species were unsuitable for dinosaurs or poisoned them.
- Climate change: too hot, too cold, too wet, too dry.
- Atmospheric change: high levels of oxygen, low levels of carbon dioxide.
- Oceanic and topographic change: Seas retreated (we know that during the late Cretaceous they did); or large volumes of fresh, cold Arctic Ocean water spilled into the Atlantic, lowering temperatures and causing drought.
- Volcanism, whose resulting soot and ash could have had the same lethal effects as predicted for impact.
- Extraterrestrial events such as supernovae explosion and meteorite impact.

Let us put these theories up against the criteria deduced by David Raup from his career-long study of extinction: "For geographically widespread species, extinction is likely only if the killing stress is one so rare as to be beyond the experience of the species, and thus outside the reach of natural selection."[32] But how widespread is widespread? Raup answered this question at Snowbird I, concluding that: "Modern biogeography is too robust for mass extinction to result from annihilation of life in a single region. . . . A global or near global crisis or environmental deterioration is required."[33]

Let me rephrase Raup's two prerequisites:

1. For a species that lives over a wide area to be driven to extinction, the cause of death not only has to be powerful, it must also be outside the experience of the species—not of individuals but species. This means that the cause must be so rare as to appear no more often than once every few hundred thousand or few million years.
2. The extinction of over 50 percent of all living species—a mass extinction—requires killing on a global scale; mass death in a region or two will not do the job.

Which of the theories summarized by Benton are rapid enough in their action to be beyond the reach of natural selection (or migration) and are also global in their reach? Medical problems, competition, and floral changes are the stuff of natural selection; they also tend to be regional. Climatic, atmospheric, and oceanic changes are widespread and therefore appealing, but they tend to occur on geologic time scales. Rather than causing mass extinctions, these gradual changes would give organisms opportunity to evolve or to migrate in response.

Changes in sea level are worth a special look, for they are the most-cited cause of dinosaur extinction. Consider the most recent dramatic change in sea level, the rise that occurred when the last glacial ice melted. The earth has a fixed amount of water during any one period of geologic time; the more that is locked up in ice, the less that is available to fill the oceans. Thus as glaciers form, sea level drops; when glaciers melt, sea level rises. During the Pleistocene Ice Ages, which ended (or at least paused) 15,000 years ago, sea level first fell and then, when much of the ice melted, rose by hundreds of meters. Although many large mammals became extinct, no mass extinction resulted. Indeed, sea level has risen and fallen throughout the history of the earth as glaciers have waxed and waned, sea floors have spread, continents have collided, and oceans have opened and closed. The record is shown by the so-called Vail curve of sea level, developed by researchers at Exxon (Figure 22).

Some of the major shifts in sea level occurred near in time to major geologic boundaries and extinctions, but many did not. The much-touted change in sea level during the Cretaceous actually

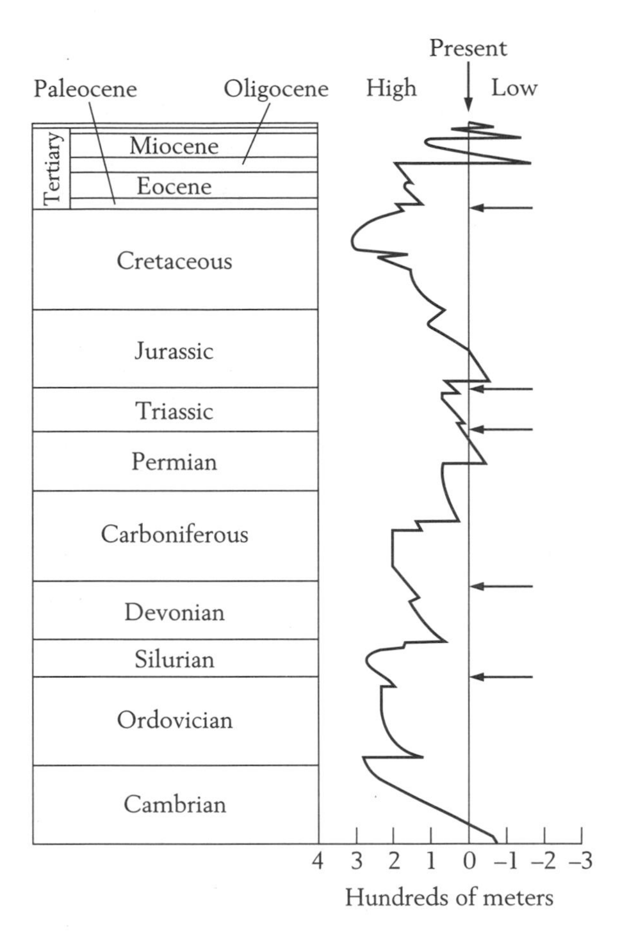

FIGURE 22 The rise and fall of sea level over the last 600 million years. Many abrupt changes fail to coincide with major extinctions and geologic boundaries. The levels of the Big Five extinctions are indicated by arrows. [The Vail curve, after Raup.[34]]

began near the middle of the period. During the long run of the dinosaurs, from *Eoraptor* in the Triassic 230 million years ago to the last *T. rex* at the end of the Cretaceous, many changes in sea level, some up, some down, many larger and quicker than the K–T change, were all nevertheless survived. In any event, let us remember that the dinosaurs lived on land. A drop in sea level, which some vertebrate paleontologists propose to explain their demise, would by definition open up more land surface on which the dinosaurs could live. The claim that such a drop caused the extinction of creatures that had lived for 160 million years appears to be contrary to logic.

David Fastovsky, co-author of an excellent book, *The Evolution and Extinction of the Dinosaurs*,[35] and Peter Sheehan, a paleontologist with the Milwaukee Public Museum, put it this way at the third Snowbird conference: "It is counterintuitive to posit that an increase in land surface area (as occurs by definition as the result of a drop in sea level) will be accompanied by habitat fragmentation [claimed by

Archibald to be a result of a drop in sea level[36]; why should the terrestrial realm be 'fragmented' by increasing the habitable area? An increase in land should provide opportunities and space not previously available to land-dwelling organisms."[37] Peter Ward has said, "We just do not know how a regression [drop in sea level] could kill anything."[38]

Of the theories on Benton's list, only massive volcanic activity and extraterrestrial events meet Raup's criteria of being global, infrequent, and lethal. Dinosaur specialist Dale Russell, reviewing the evidence in 1979, thought volcanism was unlikely to provide the answer because it tends to be gradual and episodic, rather than sudden like the K–T extinction.[39] In Chapter 6 we saw how the details of the Deccan eruptions fail to corroborate the volcanic alternative. Now it is time to turn from theory to the dinosaur fossil record.

DINOSAUR FOSSIL EVIDENCE

We wish to learn how the dinosaurs died, not how they lived; therefore our interest is in their last few million years during the late Cretaceous. How many dinosaur species were living then, where are their remains, and what do they tell us? According to paleontologist Peter Dodson, only about 2,100 articulated bones of dinosaurs have been collected, and they span 160 million years.[40] If spread evenly, we would have one specimen for each 75,000 years, but in fact the discovered remains of dinosaurs are highly clustered in time and space. All the *Tyrannosaurus rex* specimens, for example, come from Montana and the Dakotas. Therefore in the rock record there are spans of millions of years during which we know dinosaurs lived but of which we have no trace. Dodson reported that 336 recognized dinosaur species have been identified, but that nearly 50 percent are known only from a single fossil specimen. The 336 species belong to 285 genera (remember that genus—plural, genera—is the next taxonomic grouping above the species level), of which over 70 percent occur only in one rock formation. Paleontologists have learned that a typical genus has several species and that the species:genus ratio therefore is usually well above 1:1. To find it so close to 1:1 (336:285) for the dinosaurs indicates that sampling has barely scratched the surface. Recent experience confirms this conclusion, for new dinosaur discoveries seem to pop up in the press every few months. Surely many more dinosaur genera await discovery. Dale Russell estimated that we have found only about 25 percent of the genera that lived during the late Cretaceous alone, which is by far the most studied period.[41]

Although dinosaur specimens are few, they occur in late Cretaceous rocks on every continent and at dozens of sites around the world. To study their extinction, all we need are dinosaur-bearing sections that extend from the late Cretaceous up to the K–T and at least a short distance beyond it into the Tertiary. At how many places in the world can such sections be found? The answer, shockingly, is *three:* Alberta, Wyoming, and Montana. The Hell Creek formation near Glendive, Montana, is by far the best studied. Dinosaur research is continuing in other countries today, in Argentina and in China, for example, and in time more sites will meet the criteria. But up to now, to provide the litmus test for dinosaur extinction theories, paleontologists have had no alternative but to rely on fossils from the upper Great Plains, and from the Hell Creek formation in particular.

What, then, do we know about the dinosaurs from the Great Plains? At Snowbird I, Tom Schopf pointed out that our knowledge of the Maastrichtian (latest Cretaceous) dinosaurs derives from only 16 known species, which have in turn been identified from only 200 individual specimens.[42] No dinosaur has captured our imagination better than the horrific *Tyrannosaurus rex,* yet only a handful of complete skulls has ever been found.

When this paucity of hard evidence is added to the problems of gaps, migration, Signor-Lipps effect, dissolution, location of boundaries, channel cutting, reworking, and so on, it is clear that any confidently definitive statement about the demise of the dinosaurs based on scarce fossil evidence is apt to be wrong. A bang can easily be mistaken for a whimper, and vice versa.

TO HELL CREEK AND BACK

The Hell Creek formation achieved notoriety in 1902 when famed dinosaur hunter Barnum Brown discovered there the first, magnificent *Tyrannosaurus rex*. Over the more than nine decades since, fossil hunters and serious paleontologists have returned again and again to northeast Montana and Hell Creek to collect and to decipher.

How did the Hell Creek rocks come to be? In late Cretaceous times, a shallow sea stretched across western North America from Canada to Mexico. As it shrank and retreated, the sea left behind a sequence of sediments formed in different marine and nearshore environments. Subsequent deposits entombed these older sediments, which were lithified (turned into rock) by the resulting heat and pressure; erosion later exhumed and exposed them for us to see today. By studying the composition and features of sedimentary rocks and their fossils, sedimentologists and paleontologists can reconstruct the history of ancient landscapes in great detail. They know that the oldest formation in the Hell Creek area, the Bearpaw Shale, formed from muds deposited on the floor of the ancient sea. Above the Bearpaw lies a sandstone that formed from the beaches left behind as the seaway retreated. It in turn is overlain by two formations deposited by streams that meandered back and forth across marshy, nearshore floodplains: the Cretaceous Hell Creek formation and higher up, the Tertiary Tullock formation. Near the base of the Tullock lie several thin beds of coal that formed in the reducing conditions in the coastal swamps. Geologists had come to accept, as the K–T boundary in Montana, the lowermost of these coals, the "Z coal," in part because no dinosaur remains occur above it. (Admittedly, a certain amount of circular reasoning is going on here.) The uppermost dinosaur fossils were thought, especially by Clemens, to lie about 3 m below the Z coal and no closer, suggesting that the dinosaurs had gone extinct scores of thousands of years before the boundary.

From decades of study, vertebrate paleontologists had come to conclusions about the end of the dinosaurs that the Alvarez theory was initially unable to shake. For example, although 36 dinosaur genera occur in rocks dating some 10 million to 11 million years before the K–T boundary, those immediately below the boundary contained only about half that many genera. To most paleontologists, this was a clear indication that the dinosaurs were on the way out well before the end of the Cretaceous and that, if anything, impact had delivered only a coup de grace. Dale Russell pointed out, however, that the higher number of genera from the older rocks was a total obtained by adding together all those found at 25 locations around the world, whereas the smaller, later number had come from only the three North American sites. This suggested that the difference might be only a sampling effect. Nevertheless, by the mid-1980s, most paleontologists who had studied the Hell Creek fossils had firmly concluded that the dinosaurs had already died out some 20,000 to 80,000 years prior to the K–T boundary, before the putative arrival of any meteorite. Some paleontologists presented evidence that the dinosaurs had already started to be replaced by mammals well down in the Cretaceous, and claimed that this showed the dinosaurs had started to disappear well before the K–T boundary.[43]

In the mid-1980s, the confidence in this picture began to wane. It was shown then that the mammals occurred not in rocks of true Cretaceous age, but in Tertiary channel deposits cut down into the Cretaceous rocks (see p. 134), thus evaporating the evidence for early replacement of dinosaurs by mammals. Standing the argument on its head, others have reported finding dinosaur remains in Tertiary rocks in the Hell Creek region and claim that the dinosaurs did not become extinct at the K–T after all![44]

Further work at Hell Creek eroded the attempts at precise correlation of the rock strata there. According to Jan Smit, previous workers had placed the K–T boundary at Hell Creek between 2 m and 12 m too high (the same point Smit made for the sediments at Mimbral).[45] When a boundary is set too high, species immediately below it appear to have gone extinct earlier than they actually did. By placing the K–T boundary too high, according to Smit, the false impression was created that the dinosaurs had gradually disappeared and had been replaced by mammals well before the end of Cretaceous time. The Z coal beds, thought to mark the bottom of the Tertiary, were shown not to be of the same age at different locations, eliminating their usefulness as a time marker.[46,47] Finally, the Hell Creek strata were found to be riddled with gaps. At the 1995 meeting of the Geological Society of America, J. K. Rigby, Jr., of Notre Dame, reported paleomagnetic studies which showed that 300,000 to 500,000 years of the rock record were missing there.[48] The modern view is that the rocks of Hell Creek cannot be matched from one spot to another with sufficient resolution to make the precise chronology of dinosaur extinction clear.

As sometimes happens, the more a phenomenon is studied, the more questions are raised and the less confident the answers become. The Hell Creek strata now appear so complicated and full of gaps that it is hopeless to attempt to use them as the litmus test of dinosaur extinction. Some scientists realized that they needed to take an entirely different approach from the traditional, one that provided large enough samples so that statistics could be employed, and which did not depend on the precise location of the K–T boundary at Hell Creek. This view gave rise to two important studies.

Survival Across the K–T Boundary at Hell Creek

The Museum of Paleontology at Berkeley contains more than 150,000 curated specimens of nonmarine vertebrate fossils—mammals, dinosaurs, turtles, snakes, and so on—from the Hell Creek and

Tullock formations. David Archibald and Laurie Bryant (both former students of Clemens) took upon themselves the task of sorting through this enormous database to count the number of species that were present both in the Hell Creek formation and in the overlying Tertiary rocks of the Tullock.[49] They found that of the 111 species of land-dwelling vertebrates present in the Hell Creek, 35 survived the K–T boundary. This was a percentage survival rate of 32 percent, meaning that 68 percent had become extinct. None of the 20 species of dinosaurs had made it; their percentage survival rate is 0. But what about the mammals? The complete extinction of the fascinating dinosaurs has obscured the devastating blow struck to the smaller and less interesting mammals at the K–T—the Archibald-Bryant study shows that only 1 out of 28 mammal species survived. How lucky we are!

Sheehan and Fastovsky divided the Archibald-Bryant database into those that lived on land and those that lived in freshwater.[50] They found that whereas 88 percent of the land dwellers became extinct at the K–T, only 10 percent of the freshwater assemblage did so. What could explain this distinct difference in survival rate? According to Sheehan and Fastovsky it arose because the dinosaurs and other land dwellers were at the top of a food chain based on living plants, some 80 percent of which became extinct, whereas the water dwellers were part of a chain more dependent on organic detritus left behind in lakes, streams, soil, and rotting logs. Our mammalian ancestors may have survived because they were part of, or were able to become part of, the detritus-based food chain.

Triumph of the Volunteers

America is a nation of volunteers. Management expert Peter Drucker has pointed out that one out of every two adults is a volunteer; when they are included, the nonprofit sector is the nation's largest "employer." Without their corps of stalwart, dedicated, (and unpaid), volunteers, the museums of America (and most of the rest of the nonprofit sector) would be but pale shadows of themselves.

Sheehan and his colleagues at the Milwaukee Public Museum realized that the dinosaur specimens in museum and university collections had been amassed not to solve the puzzle of extinction, but for a variety of other reasons—typically because the specimens were large, or rare, or unusual in some way. They reasoned that in order to get a better handle on the question of dinosaur extinction, a new, large sample collected specifically for that purpose was needed. If enough fossils from the upper Cretaceous could be collected, scientists might be able to tell whether the dinosaurs were already declining well before the boundary, as the anti-impactors claimed, or whether they were found right up to the boundary. At least they might be able to shrink the ghastly gap.

But who would do all the work of collecting the required large number of new dinosaur specimens? To the museum professionals, the answer came at once—the volunteers! Sheehan spent three summers in North Dakota and Montana collecting in the Hell Creek formation, accompanied each time by 16 to 25 carefully trained and closely supervised volunteers from the "Dig a Dinosaur" program of the Milwaukee Public Museum, who paid $800 for the privilege. They spread out in "search parties," scouring the Hell Creek terrain for any sign of a dinosaur fossil. When a volunteer found a specimen, a paleontologist went over to make the identification, which was then logged into the computer. Almost all specimens were left in place rather than being collected and removed. To reduce the effect of different sedimentary environments, the collectors restricted their efforts to one of three sedimentary facies (distinct rock types that sedimentologists can identify). The volunteer workers logged the amazing total of 15,000 hours of careful fieldwork and found 2,500 dinosaur fossils. A key point of the study, one that differs from earlier work, was that they recorded not only whether a given species persisted at a certain level, but how many times it occurred. In other words, they measured not only taxonomic diversity (how many species are present no matter how rare), but what they called ecological diversity (how many individuals are present). This is a distinction with a difference: A species that was almost but not quite eliminated would leave taxonomic (naming) diversity unchanged—only a single individual would retain the taxon's name on the list. On the other hand, ecological diversity (number of individuals) would have plummeted, showing that it is obviously the more informative measure for tracing patterns of extinction.

The Milwaukee crew divided the Hell Creek formation into three units of approximately equal thickness, with the top one reaching up to the K–T boundary, and measured the number of dinosaur families in each third. Their most diligent search found dinosaur fossils within 60 cm of the K–T boundary, thus shrinking the ghastly gap over which Luis Alvarez and Clemens had tangled far into the Berkeley night. Their focus on ecological diversity allowed them to conclude: "Because there is no significant change between the lower, middle, and upper thirds of the formation, we reject the hypothesis that the dinosaurian part of the ecosystem was deteriorating during the latest Cretaceous. These findings are consistent with an abrupt extinction scenario."[51]

Not surprisingly, Clemens and Archibald, among others, disagreed. Clemens, for example, citing studies by Peter Dodson,[52] which he said showed a decline in diversity, held firm: "Any viable hypothesis of the causal factors of dinosaurian extinction must account for the evidence of decrease in generic diversity."[53] Sheehan and Fastovsky countered by also quoting Dodson: "There is nothing to suggest that dinosaurs in the . . . Maastrichtian were a group that had passed its prime and were in a state of decline."[54]

The work of the good folks from the Milwaukee Public Museum drives a nail in the coffin of arguments for the gradual decline of the dinosaurs. In an interview published in 1994, looking back, Clemens appeared to doubt the earlier evidence: "The 'Ghastly blank,' the unfossiliferous meter or so separating the stratigraphically highest dinosaurian bones and the iridium-enriched layer, might well be the product of leaching of fossils from the uppermost Hell Creek by acidic ground waters derived from the widespread Tullock Swamps."[55]

Reviewing this debate, a territorial chauvinism becomes obvious. Dinosaurs lived on every continent, yet the entire argument about their extinction is based on evidence from one small area in eastern Montana where, as Dale Russell pointed out in his review of Archibald's book (*Dinosaur Extinction and the End of an Era*[56]), "tabulations of dinosaur species are based on about 100 incomplete skeletons."[57] On the basis of this limited sample, from a geologically complicated and minute fraction of dinosaur-land, some paleontologists have made the most categorical statements about dinosaur extinction. Yet surely, as we saw at Zumaya, local conditions can cause a blank. Perhaps for unknown reasons, the dinosaurs simply left that part of the Hell Creek area. In the Raton formation in Colorado, the tracks of duck-billed hadrosaurs are found only 37 cm below the iridium layer.[58] Since tracks are not reworked and required a living, breathing dinosaur, this proves that the dinosaurs lived to within a few thousand years of the K–T boundary. In Mongolia, more dinosaur species are found in the latest Cretaceous than below it.[59] In China, dinosaur fossils are found so close to the K–T boundary that some say they actually transcend it.[60] In the Deccan, dinosaur eggshells are found in an intertrappean bed just at the boundary.[61] All this evidence shows that the dinosaurs did not go extinct well before the K–T boundary, but lived right up to it. If a blank does exist, it is not so ghastly after all, but merely, once again, an artifact of the Signor-Lipps effect.

In Chapter 9 I listed the two major predictions for the fossil record made by the Alvarez theory. How well has the dinosaur evidence met them?

PREDICTION 1: Prior to the K–T boundary, the dinosaurs were not already going extinct for some other reason. Their extinction was sudden and right at the boundary.

Dinosaur expert Peter Dodson, and the work of Sheehan and colleagues, indicate that the first part of this prediction is met: There was no gradual decline. The dinosaurs did not become extinct well before the K–T boundary, but lived right up to it.

PREDICTION 2: Dinosaur fossils are not found above the iridium horizon.

With a few unconfirmed exceptions, this prediction is also met. The "Tertiary dinosaurs" from Hell Creek and China may not be that at all, but instead result from misplacement of the K–T boundary or from reworking. If it does turn out that a few dinosaurs survived into the Tertiary, they will not be sufficiently common to falsify this prediction or to figure importantly in earth history.

In addition to these two, there is a question that can be asked, even if it does not lend itself to a third prediction: Can the Alvarez theory help to explain the selectivity of the K–T extinction? Since the days of Baron Cuvier, the French father of taxonomy and paleontology, at the turn of the eighteenth century, the extinction at the end of the Cretaceous has been known to have been strangely selective. Marine reptiles, flying reptiles (including the pterodactyl, named by Cuvier), and the dinosaurs died out, as did many marine invertebrates, including the ammonites and most of the planktonic foraminifera. But many terrestrial vertebrates—snakes, crocodiles, turtles, and mammals—and some plants, survived. The impact-extinction theory ought to make it easier to explain this peculiar pattern. On the other hand, if the theory does not help, it may not be because it is wrong but rather because we lack knowledge and imagination. After all, no one has yet been able to explain under *any* theory why the crocodiles and turtles survived and the dinosaurs did not. Were the impact theory also to prove wanting, it would be no worse off than the theories that geologists have traditionally preferred—those theories cannot explain the selectivity either. But certainly, an ability to explain the selectivity of the K–T extinctions would immeasurably strengthen the Alvarez theory.

HELL ON EARTH

To try to understand whether and how the Alvarez theory might help to explain the selectivity of the K–T extinction, we need to know what would happen when a 10-km to 15-km meteorite strikes the earth. The two halves of the Alvarez theory—that impact occurred and that it caused the mass extinction—are linked by the assumption that the resulting effects would be sufficiently lethal to cause the death of 70 percent of all species. The Alvarez team had precious little evidence for this assumption; indeed, to gain some idea of the effect of a global dust cloud, Luis had to rely on the century-old Krakatoa report of the Royal Society. But over the last couple of decades, the science of computer modeling of impact explosions has made great strides, such that it is now possible to say more confidently what the actual effects would be, though not how they would all interact with each other and with living organisms. (The discussion in the next dozen paragraphs is taken largely from the work of modeler Brian Toon and his colleagues.[62]) To take the subject from theory to practice, in July 1994 the entire world saw an actual planetary impact when the fragments of Comet Shoemaker–Levy 9, some estimated to be 2 km in diameter, collided with Jupiter. To the delight of Gene Shoemaker, the effects were even more spectacular than the impact modelers had predicted.

According to the Alvarez theory, 65 million years ago a comet or asteroid 10 km to 15 km in diameter approached the earth (we do not know which it was, but either would have had the effects I am about to describe). It was traveling at cosmic speeds somewhere between 20 km and 70 km per second and for that reason carried with it an energy on the order of 10^{31} ergs, or 100 million megatons of TNT (100,000,000,000,000 tons of TNT), far more energy than contained in all the world's nuclear weapons at the height of the Cold War. Once the object struck, that amount of energy had to be dissipated. An almost irresistible force was about to meet an immovable object.

As we saw when Shoemaker–Levy 9 struck Jupiter, a meteorite entering a planetary atmosphere at cosmic velocities generates a giant shock wave—a kind of cosmic backfire—that sends a 20,000-degree jet of flame thousands of kilometers back up the incoming trajectory. In the largest impacts, the entire atmosphere in the vicinity of the entry point is blasted into space.

The midair explosion of a meteorite at Tunguska in Siberia in 1908 and the eruption of Mount St. Helens in 1980 were strong enough to level trees for miles around. The K–T impact event released an amount of energy millions of times greater than these relative pip-squeaks. The resulting shock wave leveled everything standing within thousands of kilometers of ground zero, providing fuel for the subsequent fires.

Sixty-five million years ago, the Yucatán Peninsula was an area of shallow sea, so that the meteorite probably landed in less than 100 m of seawater. Modeling indicates that the resulting earthquake caused submarine landslides that displaced huge volumes of seawater and generated a tidal wave that dwarfed even the most devastating in human history. Traveling outward at about 0.5 km/sec, like the ripples from a stone hurled by a giant, this ancient tsunami rose to a height of 100 m and rolled inexorably across the oceans. Hardly slowing as it went ashore, it traveled inland for 20 km, inundating the coastal plains on half the globe.

As the meteorite penetrated deeper into the earth, a huge shock wave converted it and the rock underneath into vapor and ejected them outward at ballistic velocities. Some 100 km^3 of excavated rock and 10^{14} tons of vaporized comet or asteroid rose to altitudes as high as 100 km. Much of this debris quickly fell back to earth, but 10 percent to 20 percent of it remained at high altitudes for months. The temperature at ground zero rose to hundreds of thousands of degrees, causing everything within a radius of several hundred kilometers to burst into flame. The expanding fireball rose quickly and within only a few hours had distributed itself around the earth. Meanwhile, the shock wave had excavated a crater 15 km to 20 km deep and at least 170 km in diameter. The impact generated an earthquake of magnitude 12 to 13, a temblor at least 1,000 times larger than any humans have ever experienced. Even 1,000 km from ground zero, the earth's surface heaved in waves hundreds of meters high.

A few minutes later, the mixture of vaporized meteorite and rock, still traveling at ballistic velocities of 5 km/sec to 10 km/sec, began to reenter the atmosphere. The individual globules were traveling so fast that they ignited, producing a literal rain of fire. Over the entire globe, successively later the greater the distance from the target, the lower atmosphere burst into a wall of flame, igniting everything below. The effect was like "a domestic oven set at 'broil'."[63] Everything that could burn did.

Smoke and soot rose to mingle with the huge number of fine particles that the explosion had carried into the stratosphere. Together they darkened the earth enough to cause the average global temperature to fall to the freezing point. Darkness came at noon, and remained for months. Photosynthesis halted and the food chain that depended upon it ceased to function.

The blast wave acted as a chemical catalyst, causing atoms of oxygen and nitrogen to combine to form various noxious compounds, many found in today's smog. Sulfur oxides joined them, for in a coincidence unfortunate for life at the end of the Cretaceous, the Yucatán rocks at ground zero included sulfate deposits. As happened in the modern eruptions of Pinatubo and El Chichón, sulfur dioxide formed tiny droplets that further obscured the sun and lowered visibility even more. Kevin Pope, Kevin Baines, and Adriana Ocampo have calculated that the impact into the sulfur-rich deposits of the Yucatán would have produced over 200 billion tons of both sulfur dioxide and of water, leading to a decade-long impact winter.[64]

As precipitation washed out the nitrogen and sulfur compounds, it generated acid rain that may have destroyed the remaining susceptible plants. Gregory Retallack of the University of Oregon[65] has found evidence in the boundary clay in Montana of severe acid leaching, possibly enough to have dispersed the iridium and dissolved the shocked minerals and spherules. Thus, Retallack says, some impacts might be "self-cleaning," eliminating traces of their own existence. Because some soils naturally buffer acids and others do not, acid rain might also explain some of the K–T extinction selectivity. For example, the floodplains of ancient Montana would have remained above a pH of 4, which according to Retallack would spare small mammals, amphibians, and fish, but harm plants, nonmarine mollusks, and dinosaurs. Acid-vulnerable plants such as the broadleaf evergreens would have suffered, whereas the acid-tolerating plants would have done better, more or less consistent with the evidence.

The rain may have acidified the surface layers of the oceans sufficiently to kill the surface-dwelling plankton and phytoplankton, which would have caused a breakdown of the oceanic food chain that was based upon them. The reactions that formed nitrogen oxides also absorbed ozone, reducing the earth's protective ozone layer and allowing ultraviolet radiation to penetrate to the surface, causing further loss of life.

Some of the vast amount of water vapor that was blasted into the atmosphere froze; the rest formed a vapor cloud that lasted for years. There it was joined by the most insidious long-term effect of the impact—a worldwide cover of carbon dioxide, generated by impact into the thick limestones (calcium carbonate) that also were present in the Yucatán Peninsula of that day. Just when the dust, smoke, and soot had dissipated and conditions might have returned to near normal, this gas cloud produced a greenhouse effect that lasted for a thousand years or more. Those creatures that had miraculously survived all that came before, now faced a millennium of greenhouse temperatures. (Recent modeling by Pope and Ocampo, however, indicates that the greenhouse effect might not have been this strong.)

Obliterating shock waves, stupendous earthquakes, enormous tsunami, a rain of fire, smoke, soot, darkness, a global deep freeze, worldwide acid rain, ozone loss, greenhouse warming—it seems a miracle that anything could have survived, and yet, remember our thought experiment on just how difficult it is to exterminate an entire species. Over 99.99% of individuals can die and enough breeding pairs might be left alive to allow the species to survive. But certainly no one can claim that the impact of a 10-km meteorite in the Yucatán Peninsula 65 million years ago lacked the power to cause the K–T mass extinction.

We know that a 100-million-megaton impact happened at K–T time; we know that it must have had some combination of the effects just described. What we do not know is just how the many lethal possibilities would have interacted with each other and with living organisms. These questions will occupy impact modelers, geochemists, paleontologists, and others, for years. Meanwhile, some paleontologists, though now objectively required to admit that impact happened, remain unwilling to grant that it had anything to do with extinction until "precise biological/ecological mechanisms are proposed that uniquely account for observed taxic patterns and the stratigraphic timing of K–T extinction and survivorship."[66] The clear implication is that the burden of proof still rests entirely with the pro-impactors: They must explain how the impact effects killed certain species and spared others. But the existence of the Chicxulub crater shifts the burden. Since we know that impact occurred, those who deny that it caused the mass extinction have just as much of an obligation to explain how species escaped as those who support the link between impact and extinction do to explain how they did not.

THE SECOND HURDLE

Looking back at the evidence described in this and earlier chapters, we can see that the predictions of the impact half of the Alvarez theory hold up well. To cap it off, geologists have located the impact crater. Further, not only do ammonites, plants, and forams offer corroborative evidence that impact led to extinction, so do the dinosaurs. Keeping in mind that theories are not proven, only disproven, I believe that it is fair to say that the core of the Alvarez theory—that impact occurred and that it caused the great K–T mass extinction—has been corroborated. It has met many tests and failed none.

Charles Officer and Jake Page do not agree. On the contrary, in their 1996 book, they write that the theory has "collapsed under the weight of accumulated geologic and other evidence."[67] In the final chapter of *The Great Dinosaur Extinction Controversy,* they cite the Alvarez theory as an example of "degenerative science," comparable in its failure to Marxism. By misleading us into searching the heavens for incoming meteorites, the authors claim, when instead we should be solving current environmental problems right here on earth, the theory is downright dangerous. But other geologists have moved in the opposite direction: They are fascinated by the possibility that the Chicxulub impact was not the only one to cause a mass extinction.

Part IV

The Transformation of Geology

CHAPTER 11

ARE ALL MASS EXTINCTIONS CAUSED BY COLLISION?

All five major mass extinctions would turn out to have been caused by the same mechanism, an asteroid collision.[1]
Luis Alvarez

In his 1983 talk at the National Academy of Sciences, Luis Alvarez left no doubt just how far he thought his theory extended. He did acknowledge that the prediction quoted in the epigraph had not yet been confirmed, but made it clear that he believed it eventually would be. If it were confirmed, then not only might the explanation of mass extinctions have been discovered, so might the driving force behind evolution itself. Though geology offers no Nobel prize, the discovery of a robust general theory for mass extinctions, and especially of one linking them to extraterrestrial causes, surely would rank as one of the great scientific accomplishments of the twentieth century and place its authors in Nobel territory.

Physicists, even more than other scientists, seek the explanation of more than individual phenomena—they want to uncover the grand unified theory that will explain how each of the fundamental physical forces arises and interacts. Einstein, for example, tried to show that both electromagnetism and gravity derived from the same fundamental "force field." He was unable to do so, nor has anyone been able to since. Few geologists have even tried to imagine an all-encompassing theory for the earth: What could the unifier possibly be for the complex and seemingly random set of processes that characterizes our planet? Yet, we must ask the question: What of impact? It first created the inner planets through accretion, then destroyed and reshaped their surfaces. It may have carved the moon from the earth. It produced the most energetic event in the last 600 million years of earth history, one that led directly to the K–T mass extinction. Even though it is counterintuitive, our intellect forces us to recognize that impact has happened thousands, indeed tens of thousands of times, since the earth cooled (though few impacts would have been the size of Chicxulub). Could the energy released by myriad impacts throughout geologic time be the grand unifier of geology?

Recognizing Impact

Before we get too far out on a limb of speculation, let us ask first whether there is hard evidence for impact at any mass extinction horizon other than the K–T. Do any others show an iridium spike, shocked minerals, and spherules, not to mention spinel, diamonds, and soot? Do any others have an impact crater of corresponding age? If the answer to these questions is no, we would have to set aside the notion that impact, beyond its singular occurrence at the K–T boundary, has played an important role in earth history. Luis's prediction would have failed.

Of course, to be prepared to base a judgment on hard evidence presumes that the indicators of impact, if once present, would remain around to be discovered and that they could be detected. Are these fair presumptions? Not really. Recall that geologic boundaries were defined, well over a century ago, primarily because they were easy to spot in the field—they tend to be places where one rock type abruptly gives way to another. But these are the very places where erosion has done its work. Almost by definition then, geologic boundaries are apt to be the location of gaps in the rocks: levels at which erosion has removed whatever was present, including any thin impact ejecta layers.

Another difficulty is that subduction has removed oceanic crust older than about 125 million years. Any extinction older than that, which includes four of the Big Five, cannot be found preserved in cores drilled from the oceanic sedimentary layer, which offers the most continuous and least disturbed sections. Instead we must seek these older boundaries in continental rocks, where erosion is more likely to have removed them.

What about our old friend iridium? If found in high concentrations it is as good an indicator as ever, but the converse is not true: Low iridium levels do not necessarily rule out impact. First, comets, which travel through space at around 45 km/sec to 60 km/sec, cause a significant fraction of all impacts (see Table 1, page 51). Asteroids move at slower velocities, averaging about 20 km/sec. Since kinetic energy is proportional to velocity squared, this three-fold difference in speed means that a crater of a given size can be produced by a comet one-ninth the size of the asteroid required to produce that same crater. Thus an impact crater produced by a comet would leave no more than one-ninth the iridium to be found in a crater of the same size formed by an asteroid. But even this is an upper limit. Comets, being as much as 50 percent ice, carry much less iridium to start with (calculations that combine crater size and composition show that the amount of iridium left by a comet might be as low as 1 percent of that left by an asteroid, too little to be detected). As Table 1 shows, larger craters are successively more apt to have been formed by comets (one of the reason most specialists now believe that Chicxulub was formed by the impact of a comet). Thus it is ironic but true that the larger the crater, the less likely it is to leave iridium behind. To carry matters further, in the largest impacts, regardless of impactor type, almost all the ejecta is blasted back out into space, escaping the earth's gravity field altogether and leaving no trace behind.[2] Still another complication is that in smaller impacts, extraterrestrial material composes only about 10 percent of the ejecta, so that if the impactor happened to be an asteroid relatively low in iridium, which some are, little would be left to find. Finally, even when iridium was present initially, reworking and bioturbation could have smeared it out, or acid leaching could have removed it. All in all, iridium is a kind of one-way indicator: Its presence is strong evidence of impact; its absence is not evidence of no impact.

Impact by either comets or asteroids, however, would leave behind shocked minerals and possibly spherules, maybe even spinel and diamond. Since these markers are less subject to alteration or removal by chemical and geologic processes, they make a better bet as indicators of impact than iridium. Most geologists continue to be most impressed by shocked quartz, the indicator that they discovered.

Finding a crater that dates to the time of a geologic boundary is fraught with the same difficulties that we encountered in the search for Chicxulub. Erosion will have erased most impact craters; others will have disappeared down subduction zones. The older the crater, the more likely one of these fates. Most crater ages are not known with precision, making it difficult to assign them to a given geologic boundary with much confidence. The Manson crater in Iowa is a good example. For years its age was known only roughly, then the first measurement gave 65 million years, and finally more precise methods yielded an age of 73.8 million years. And not only must we know the age of a candidate crater, we must know the age of the extinction boundary with which it might be correlated. But the ages of many boundaries, and even their positions, have yet to be pinned down.

When we consider all these uncertainties, the accidental finding of the iridium spike at Gubbio, and the diligent search that led to the discovery of the iridium-rich layer amid the lava flows and intertrappean sediments of the Deccan, appear all the more remarkable. Even if impact has occurred at another geologic boundary, we could easily miss it. If after a diligent search, however, no evidence of impact has turned up, practical geologists, with limited time and resources, would move on to fields with more chance of results. While the absence of evidence may not be evidence of absence, it is discouraging. Being human, scientists tend to go where positive evidence and rewards can be found.

THE BIG FIVE

The K–T mass extinction was one of five in which more than 70 percent of species died. If there is anything to the notion that impact has caused other mass extinctions, it is here, among the other four, that we should first look. Table 4 summarizes the ages of the Big Five plus the Eocene–Oligocene and Jurassic–Cretaceous extinction boundaries, and the evidence of impact that has so far been found associated with each. The three right-most columns give age and size information for craters that happen to have the same approximate age as the boundary. The table implicitly asks for each of these extinctions: Is there any evidence of impact, and is there a large crater of the same age?

THE LATE DEVONIAN

The earth guards its secrets. Each of the Big Five extinctions, when examined in detail, turns out to be complicated and different. Take, for example, the late Devonian extinction. The iridium there, not high to begin with, appears to be strongly associated with the remains of the bacterium *Frutexites*. Some experts believe that *Frutexites* was able to extract and concentrate iridium from seawater, indicating, they say, that iridium is not a reliable marker of impact after all. Pro-impactors respond that *Frutexites* was able to concentrate iridium just because an impact inserted excess iridium into the oceans in the first place.

TABLE 4

Evidence for Impact at the Big Five, Eocene–Oligocene, and Jurassic–Cretaceous Extinction Boundaries

Extinction boundary	Age (m.y.)	Evidence of impact	Crater?	Age (m.y.)	Size (km)
Eocene–Oligocene	33.7 ±0.5	Tektites, microtektites, shocked quartz, coesite	Popigai, Siberia; Chesapeake Bay, U.S.	35.7±0.2 35.5±0.6	100 85
Cretaceous–Tertiary	65.0	High iridium, shocked minerals, microtektites, spherules, spinel, diamond	Chicxulub, Mexico	65.0	170–300
Jurassic–Cretaceous	~145	High iridium, shocked quartz	Morokweng, South Africa	145	70–340
Triassic–Jurassic	~202	Shocked quartz, weak iridium peak, fern spike	Manicouagan, Canada; Puchezh-Katunki, Russia	214±1 220±10	100 80
Permian–Triassic	~250	Weak iridium peak, shocked quartz, microspherules reported	Araguinha, Brazil	247±5.5	40
Late Devonian	367	Microtektites, weak iridium peak	Siljan, Sweden	368±1	52
Ordovician–Silurian	~438	Weak iridium peak	None known of this age		

When we study the geologic sections that mark the late Devonian interval in detail, we find not a single extinction boundary but several smaller ones spaced over a few million years. Several iridium-rich layers are found at this horizon. One expert, George McGhee of Rutgers, who wrote a fine book on the Late Devonian mass extinction, believes that *three* impacts occurred, and indeed, several craters do date to this part of geologic time.[3] The Siljan crater in Sweden is the most promising; it is also of interest because it was the focus of the deep-earth methane hypothesis of Thomas Gold of Cornell University. Gold convinced himself, and then the Swedish Power Board, that the impact of a large asteroid would produce fractures that would tap deep-seated reservoirs of gases, among them methane, which then might be produced in commercial quantities. The Swedish Power Board drilled the Siljan structure but found no methane.

THE JURASSIC–CRETACEOUS

Impact craters have several ways of escaping notice. They may hide beneath seafloor sediments, arctic ice, flood basalt flows, and younger sediments. Tectonic plates may carry them down to oblivion. Or, more mundanely, erosion may obliterate them. But recently yet another hiding place has been discovered.

In the southwestern part of Africa lies the great Kalahari Desert. Until the advent of four-wheeled drive vehicles, travel in the Kalahari was next to impossible as the region is covered by a layer of sand 100 m thick. In the early 1990s, aerial gravity and magnetic surveys, of the kind done in the search for oil-bearing structures in the Yucatán, revealed a nearly circular structural dome buried beneath the Kalahari sands near the South African town of Morokweng. Drills sent down into the underlying bedrock brought back melt rock containing an iridium anomaly and shocked quartz. Although at first the structure appeared to be 70 km wide, further data analysis suggests a diameter of 340 km; if so, it would be even larger than Chicxulub. Zircons from the Morokweng melt rock give an age of 145 million years.[4] The accepted age of the boundary between the Jurassic period and the overlying Cretaceous (see Figure 2, page 8), at which, according to Sepkoski, 38 percent of species became extinct, is also 145 million years.

THE PERMIAN–TRIASSIC

The "mother of mass extinctions," the Permian–Triassic, is a critical case. It marks such a vast change in the history of life that, like the K–T, it not only separates two geologic periods, but two great eras—the older Paleozoic and the younger Mesozoic (see Figure 2, page 8). But the Permian–Triassic extinction is much the larger—96 percent of species are estimated to have expired then, compared with 70 percent at the K–T. About 67 percent of reptile and amphibian families disappeared, opening the way for the rise of the dinosaurs; 33 percent of all insects, which usually survive almost anything, disappeared. Prior to the extinction, most marine organisms made their living anchored to the seafloor; those that came after tended to crawl on or to float above the bottom. This led Richard Monastersky of *Science News* to joke that we owe our modern seafood menu of "lobster bisque, fried calamari, seared tuna, and even sea urchin sushi"[5] to the evolutionary path laid open for the ancestors of these delectable creatures (well, most of them anyway) by the Permian–Triassic mass extinction.

If high iridium levels, shocked quartz, or spherules were to be found in Permian–Triassic boundary layers, the case for impact would be greatly strengthened. So far, each has been reported, but in no instance have the reports been confirmed to the satisfaction of even the pro-impactors. Iridium at the Permian–Triassic boundary, for example, appears to be a factor of 10 lower than at the K–T. At the 1996 meeting of the Geological Society of America, Greg Retallack showed photomicrographs of quartz from the Permian–Triassic in Antarctica that he claimed exhibited planar deformation features.[6] Specialists such as Glenn Izett and Bruce Bohor, however, were unconvinced.[7] Finally, there is no good candidate crater, as Araguinha (see Table 4) appears to be too small.

Douglas Erwin of the National Museum of Natural History of the Smithsonian, in his definitive account of this boundary, *The Great Paleozoic Crisis*,[8] lists theories that have been proposed to explain the Permian–Triassic mass extinction. His list is not as long as Jepsen's (page ix), but numbers 14 and includes such familiar K–T suspects as global cooling and flood basalt eruptions. The latter idea received a recent boost in a paper by Paul Renne of the Berkeley Geochronology Center and his colleagues,[9] who used the argon-argon method to date volcanic rocks of Permian–Triassic age from southern China at 250.0 ± 0.2 million years, exactly the same age his group obtained for the Siberian basalts. They propose that volcanic sulfur emitted during the Siberian eruptions caused a strong pulse of acid rain, as Retallack has argued happened at the K–T. The acid rain, together with increased concentrations of various volcanogenic poisons, caused the great Permian–Triassic mass extinction. But some geochronologists doubt that Renne's age results are quite as precise as he claims. As we saw in the case of the Deccan intertrappeans, the exact timing of events, though crucial to the argument, is exceedingly hard to pin down. Erwin has written that much of the Siberian traps eruption occurred later, in the Triassic, in which case it was too late to cause the extinction.[10]

Another group recently suggested that the culprit in the Permian–Triassic mass extinction was carbonated water, a familiar and seemingly innocuous liquid.[11] These researchers propose that in the late Permian, large amounts of carbon dioxide accumulated at the bottom of the sea. Cold seawater near the surface began to sink and displaced the abyssal gas-rich layer, which rose and released its dissolved carbon dioxide. The carbonated water thus produced caused shelled marine organisms to die off. The higher levels of carbon dioxide in the atmosphere then induced a true greenhouse effect, resulting in additional extinctions. Marrying several theories together, the scientists acknowledge that either meteorite impact or the eruption of the Siberian Traps might have triggered the turnover of the oceans and the release of carbon dioxide.

While the carbonated water theory may seem far-fetched, scientists think that something frighteningly similar took place on a smaller scale at Lake Nyos in Cameroon in 1986. One day, from this extraordinarily beautiful blue volcanic crater lake, there burst a cloud of invisible gas so deadly that it instantly killed over 1,000 people and all the cattle and other animals in the vicinity. So few people survived that it has been difficult to find eyewitnesses; the few who have been interviewed say that a fountain of water hundreds of feet high sprang without warning from the center of the lake. The culprit at Lake Nyos is believed to be carbon dioxide that accumulated at the base of the lake and then, for unknown reasons, suddenly erupted.

To account for the great Permian–Triassic mass extinction, Erwin prefers what he calls a "Murder on the Orient Express" theory. In the Agatha Christie story, a man is found murdered on the famous train—his body has not 1 but 12 knife wounds. Christie's masterful sleuth, Hercule Poirot, using only his "little grey cells," deduces that each of 11 passengers, plus the porter, stabbed the man once to avenge a terrible crime committed years earlier. Similarly, Erwin theorizes that the Permian–Triassic extinction was due not to a single cause but to several acting at once.

Early on in this book, I took the position that the principal reason the solution to the mystery of dinosaur extinction was so long in coming is that it had a single cause, impact, that was unknown to geologists until the 1970s. An event that happens once in 65 million years can hardly be the first explanation that comes to mind. But Erwin shows us that a second kind of singularity might exist. He proposes, not a single cause that is exceedingly rare and therefore unfamiliar, but a coincidence of several familiar causes. It is their coincidence at a single moment in geologic time that is rare, perhaps even unique. If Erwin is right, then the greatest mass extinction of them all was not caused by impact. On the other hand, a single confirmed grain of shocked quartz at the Permian–Triassic boundary, which Retallack says he has found, would corroborate the claim that impact had occurred then.

THE TRIASSIC–JURASSIC AND THE EOCENE–OLIGOCENE

The Triassic–Jurassic boundary in Italy contains shocked quartz and a set of weak iridium peaks; these need to be confirmed and replicated at other localities. Two scientists concluded that the plant extinction at the Triassic–Jurassic took less than 21,000 years, analogous to the sharpness of the K–T fern spike and supportive of impact.[12] On the other hand, the boundary appears to be just a little younger than each of the two candidate craters listed in Table 4. However, the Triassic–Jurassic boundary age of 202 ± 1 million years was not measured directly on something like a boundary clay, but instead is based on analysis of a volcanic rock that was interpreted by the researchers to be just younger than the boundary. It is possible that as more measurements are taken, the accepted age of the Triassic–Jurassic will shift.

At the Eocene–Oligocene boundary, approximately 34 million years ago, some 35 percent of marine genera became extinct (meaning that two or three times as many *species* did), as did many mammal species. Deep-sea drill cores show an iridium spike near the boundary, as well as the kind of spherules and shocked quartz found at the K–T. Even coesite, the high-pressure form of quartz and a diagnostic indicator of impact, is present. Not one but two craters date to the Eocene–Oligocene section of the geologic record. The Popigai structure in Siberia, 100 km in diameter, is dated at 35.7 ± 0.2 million years.[13] In 1994, Wylie Poag of the U.S. Geological Survey discovered a large crater buried beneath Chesapeake Bay that dates to 35.5 ± 0.6 million years.[14] At an estimated diameter of 85 km, Chesapeake Bay is the largest impact structure yet discovered in the United States. Its age, and the composition of the associated breccia, are consistent with Chesapeake Bay being the source of the North American tektites. Two craters the size of Popigai and Chesapeake Bay should have been created at approximately 10-million-year intervals, yet these two were struck within a few hundred thousand years of each other, showing that the laws of chance can produce seemingly improbable results.

Both craters are just older than the Eocene–Oligocene boundary, and, if all the dates are correct, could not have caused it. The age of the boundary, however, has been argued at great length in the literature, and may not be completely firm.[15] But the impact markers also appear to be just slightly older than the boundary, and no mass extinction lines up with the impact evidence.

TANTALIZING EVIDENCE

Of the seven major geologic boundaries (the Big Five, the Eocene–Oligocene, and the Jurassic–Cretaceous), only the Ordovician–Silurian lacks a sizable impact crater of approximately the same age, though in some instances the age fit is not good. Five of the six boundaries that are 200 million years old or older have craters of nearly the same age, which is remarkable considering that most craters that old have been obscured or removed by erosion. Shocked minerals or microtektites have been reported at six of the seven boundaries, though the claim for the Permian–Triassic is new and controversial. All seven have iridium concentrations that appear to be higher than background, though often by only a little.

While many of these individual pieces of evidence are weak, taken as a set they are impressive. Not enough firm evidence is available to corroborate the claim that impact is responsible for any other mass extinction boundary than the K–T, yet more than enough exists to justify a continuing investigation, providing an opportunity for earth scientists.

DID IMPACT CAUSE *ALL* EXTINCTIONS?

David Raup, always trying to see the big picture, could not restrain himself from asking a question that even he had to admit was rash and seemingly ridiculous: Could *all* extinctions of significant numbers of species—not just the major mass extinctions—have been caused by impact?[16] Since there are so many geologic boundaries to investigate, this impertinent question cannot be answered by going into the field to examine rocks and collect fossils, but perhaps it can be answered in theory. Naturally, where angels and other geologists would fear to tread, Raup rushed in. And his distinguished record shows him to be no fool.

One way to describe events spaced out in time is to refer to the *mean waiting time* between those events of a given size, as we do for the 100-year flood and the 100-year wildfire. On average, a flood the size of the 100-year flood shows up every 100 years; in practice two can occur in successive years, or none can appear for several hundred years. Raup, who calls a spade a spade, used Sepkoski's compilation of the extinction records of genera, and the concept of waiting time, to build what he called a "kill curve" (Figure 23).[17,18] It shows how much time passes on the average between extinction events of various sizes: the 1-million-year extinction, the 10-million-year extinction, the 100-million-year extinction, and so on.

Extinctions that destroy 5 percent of species occur about every million years. Interestingly, this is the approximate length of the "biostratigraphic zone," the minimum unit of geologic time that paleontologists can detect using specific assemblages of fossils. An extinction the size of the K–T has a waiting time of about 100 million years, whereas one the magnitude of the more lethal Permian–Triassic occurs at intervals of 1,000 million or even 10,000 million years. Since life has never been completely exterminated, Raup assumes that the curve must level off to form an "S," never reaching 100 percent killed, no matter how long the waiting time.

Raup next turns to Shoemaker's observations of comets and asteroids, which allow an estimate of how frequently craters of different sizes form (Table 5).[20] Figure 23 and Table 5 use two completely independent sets of data, one obtained from the record in the rocks, the other derived from searching the heavens. The first relates waiting time to percent species killed; the second relates waiting time to crater size. We could write an equation that would plot out each graph, and from high school algebra we know that we could eliminate the common variable from the two equations, waiting time, and relate percent species killed directly to crater size. In Figure 24, Raup has done so. (The curve is dashed above the 150-km-crater diameter because that is as far as Shoemaker's estimate went. Since these estimates inevitably have large associated errors, the dashed upper and

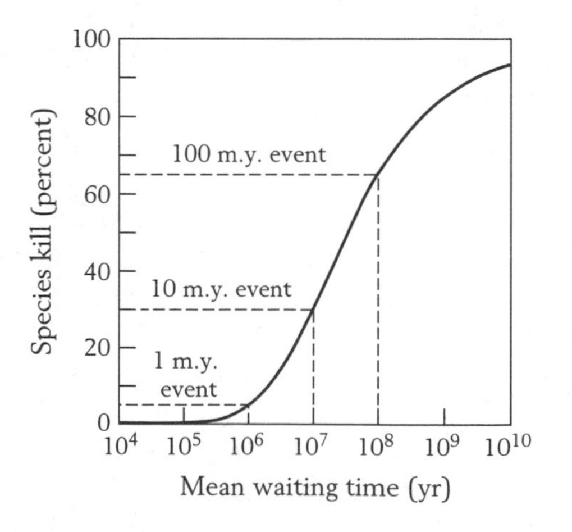

FIGURE 23 Raup's kill curve, showing the average time spacing between extinction events of different intensity. The Big Five extinctions (with the possible exception of the Permian–Triassic at 96 percent), are 100-million-year events. [After Raup.[19]]

TABLE 5

Frequency of Formation of Craters of Different Diameters

Crater diameter (km)	Mean waiting time between impacts (yrs)
> 10	110,000
> 20	400,000
> 30	1,200,000
> 50	6,200,000
> 60	12,500,000
> 100	50,000,000
> 150	100,000,000

lower curves show how far in each direction the actual curve might lie.) If we knew which craters had associated extinction events, their positions could be plotted on the impact-kill curve. As noted in the discussion of Table 4, however, the only crater about which we are certain is Chicxulub. Michael Rampino and Bruce Haggerty[21] make an educated guess that three other craters can be added to the list as shown in Table 6 and Figure 24.

Is the impact-kill curve credible on its face? It predicts that the largest extinctions are associated with craters at least 140 km in diameter. This is plausible because the largest extinctions and the largest craters each occur about every 100 million years. There have been five major extinctions since the Cambrian began, and, Shoemaker's estimates tell us, about the same number of giant impacts. Note again that these two calculations are completely independent. Sepkoski's best estimate of species killed at the K–T boundary is about 70 percent, corresponding to a crater of about 150 km. (If Chicxulub is actually 300 km in diameter, as some argue on the basis of its buried topography and gravity structure, then the actual impact-kill curve would have to be closer to the lower dashed line.) None of the points other than Chicxulub is precisely located but they do fall within the upper and lower boundaries of the curve. On the other hand, as far as we know, some sizable craters have failed to produce mass extinctions. Neither the 24-km Ries Crater in Germany, nor the 45-km Montagnais Crater, located in the seafloor off Nova Scotia, is connected to an extinction. If the reasoning behind the impact-kill curve is correct, and if Ries and Montagnais have no associated extinction, then the curve must hug 0 on the Y axis until

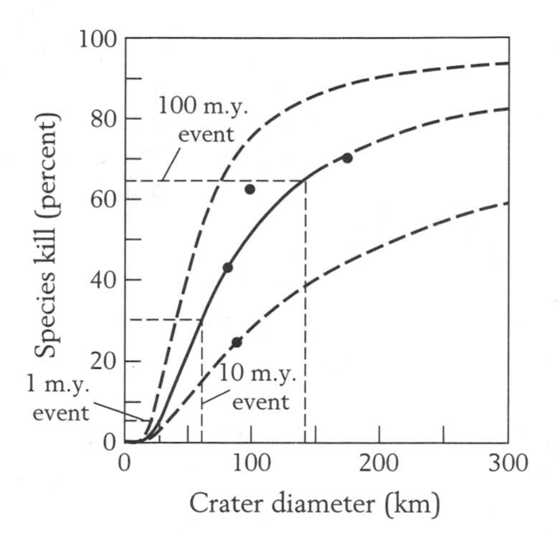

FIGURE 24 The impact-kill curve, combining Raup's kill curve (Figure 23) and Shoemaker's estimates of the frequency of formation of craters of differing sizes (Table 5). The four sites listed in Table 6 are plotted, though only Chicxulub is confirmed. [After Raup.[22]]

it reaches the point corresponding to a crater of around 45 km in diameter, when it must begin to rise steeply. This could be the case, for example, if a certain critical impactor mass were required before extinctions become global and massive.

The stage, the smallest unit into which geologists subdivide the rocks of the geologic column, represents a fundamental subdivision of earth history. In the 600 million years since the Cambrian began, Sepkoski identifies 84 stratigraphic intervals, most of them stages, giving an average duration for a stage of approximately 7 million years. If impact causes all extinction, as Raup rashly considered, then craters large enough to be associated with extinction ought to have about the same waiting time as the duration of an average stage. Is that the case? To find the answer, inspect Table 5: Note that 7 million years is the mean waiting time for a crater just over 50 km in diameter. A crater of that size releases about 5 million megatons of energy, roughly 50 million times the power of the atomic bomb that

TABLE 6

Candidate Craters for the Impact–Kill Curve

Crater	Size (km)	Percentage of species killed
Puchezh–Katunki	80	43
Chesapeake Bay	85	25
Manicouagan	100	62
Chicxulub	175	70

destroyed Hiroshima; according to Figure 24, it would result in the death of about 20 percent of species. A loss of 20 percent of species every 7 million years is equivalent to a 100 percent turnover in 35 million years, which is only about 6 percent of the time that has elapsed since the Cambrian began. Thus it is more than enough to account for the record of extinction observed in the rocks.

To answer Raup's question as he answered it: Yes, in theory, impact could have caused all extinctions. To turn the question around, since it is inescapable that the earth has been bombarded by meteorites of a range of sizes since life began over 3.5 billion years ago, and since even a modest-sized impact releases huge amounts of energy, how are we to escape the conclusion that not just in theory, but in practice, impact has caused many extinctions?

THE TEMPO OF EVOLUTION

There may be an additional way to shed light on the role of impact: by focusing not on the mass extinctions themselves, but on the normal intervals of background extinction in between. If impact drives mass extinctions, then in the times between impact events, few extinctions would be expected. Can we tell whether the tempo of extinction and evolution in between the large extinctions is consistent with a history of impact? After Darwin, evolutionists came to have the view that natural selection operates steadily, all the time. Organisms continually undergo small changes that, when summed together, produce large effects. As long as environments are stable, natural selection operates to adapt organisms ever more perfectly. When environments change, natural selection allows them to adapt just enough to keep pace. This model conforms exactly to the uniformitarian view of earth behavior: Change is gradual, but cumulative, and in time can be prodigious.

If this gradualistic view is correct, the fossil record should reflect it. As natural selection works its way, plants and animals should evolve steadily, little by little, leaving much of the work of evolution to be done in between mass extinctions. But, as Darwin's contemporaries knew, this is not really what the fossil record reveals. Instead, most evolutionary change occurs in a burst right after a new species diverges from its ancestor. After the initial spurt, species change little, sometimes remaining static for millions of years. For example, the lampshell brachiopod genus *Lingula* found today appears just like its 450-million-year-old fossil ancestor. Niles Eldredge of the American Museum of Natural History and Stephen Jay Gould called this model "punctuated equilibrium," known to aficionados as "punc eek," and when they proposed it in 1972, most evolutionists scoffed.[23,24] Today, however, many believe that not only species, but whole ecosystems, remain stable for long periods of time, until something disturbs them enough to cause a multitude of extinctions. The motor of evolution then revs up and gives rise to new species that are adapted to the postextinction conditions. Eldredge believes that "Nothing much happens in evolution without extinction first disrupting ecosystems and driving many preexisting, stable species extinct. And extinction is almost always the result of the physical environment's changing beyond the point where species can relocate by finding familiar habitat elsewhere."[25]

If punc eek is the rule, then something punctuates evolution. That something produces extinctions that in turn open up ecological niches into which the pressure of natural selection propels a new set of organisms. According to this notion, the driving force behind evolution is the punctuator itself. What could it be? If we follow where Raup, Gould, and Eldredge would lead us, we see that the punctuator must be unfamiliar to species over long periods of geologic time and must disrupt the environment beyond the ability of species to adapt or migrate. Though mass extinction may have more than one cause, however we come at the question, we find hints, if not corroboration, that impact may have played a more important role in the history of the earth than almost anyone has appreciated.

Chapter 12

Are Extinction and Cratering Periodic?

They're regularly spaced in time.[1]
David Raup

During the 1970s, when the Alvarezes were developing their theory, a young paleontologist named John Sepkoski was at work at the University of Rochester, compiling the ranges of geologic ages during which each family of fossil organisms lived. (Recall that biologists subdivide organisms into kingdom, phylum, class, order, family, genus, and species. We belong, in the same order, to the animals, chordates, mammals, primates, hominids, genus *Homo,* and species *Homo sapiens*). Sepkoski was not going to all this trouble in order to study mass extinction, but rather to learn more about how biologic diversity has changed over geologic time. Meteorite impact was the furthest thing from his mind.

Sepkoski scoured the world literature of paleontology, searching out even the most obscure journals in the most unfamiliar languages, slowly adding information to his database. The data he entered for each family were simple: name, geologic age of the oldest and youngest recorded occurrences of species belonging to the family, and the literature references. Sepkoski was fortunate to have had the encouragement of his senior colleague at Rochester, David Raup, who happened to be predisposed toward the statistical approaches to which a large database lends itself. By 1978, both scientists had moved to the University of Chicago, further strengthening a department of paleontological powerhouses. There, Sepkoski continued to upgrade and polish his compendium, until it contained 3,500 families and 30,000 genera.[2] One day a senior colleague, the late Tom Schopf, told Sepkoski that he was up for tenure consideration that year. Such an announcement, like the discovery that one is to be hanged in a fortnight, "concentrates [the] mind wonderfully."[3] In a panic to publish quickly, the young professor decided that his best bet was to try to get out an article based on his compendium. Sepkoski worked frantically for months, only to have Schopf return to say that he had been mistaken—Sepkoski's tenure would be decided the *following* year. But by this time Sepkoski had gone too far to turn back.

Raup had the idea that instead of merely perusing the compendium (a task sufficiently boring as to cause even a quantitatively minded paleontologist to nod off) it might be examined with the aid of a computer to see whether any interesting patterns emerged. Raup and Sepkoski viewed an assortment of graphical computer plots, even standing across the room to see whether a pattern recognizable only as a "gestalt" would emerge. Sepkoski suggested that it might be interesting to compute how the rate of extinction had varied through time. A few days later, Raup brought the new plot, shown in Figure 25, into Sepkoski's office. "Do you see it?" he asked, "They're regularly spaced in time."[4]

Raup had culled Sepkoski's data in the following ways: He had examined only the most recent 250 million years, when geologic ages are more precisely known; he had removed families whose ranges or identity were poorly known; and, since we have no way of knowing how long they may live, he deleted families that have not yet become extinct. He divided the 250 million years up into the 39 stratigraphic stages that geologists have recognized (geologists divide time into eons, eras, and periods; rock units into systems, series, and stages—the

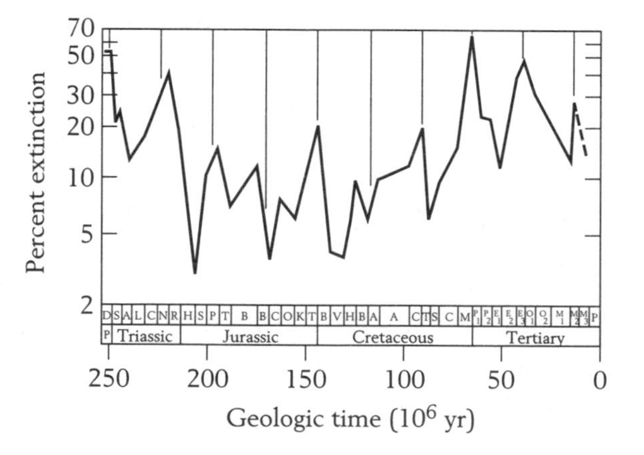

FIGURE 25 Raup and Sepkoski's 1984 plot of extinction periodicity.[5] The best-fit cycle, at 26 million years, is shown by vertical lines.

Maastrichtian, for example, is a stage in the Cretaceous system), and then plotted the percentage of extinction within each stage. Each data point came from calculating the number of families that became extinct within a stage as a percent of all the families that lived during that stage. (Removing the families that are still alive makes the denominator of this fraction smaller and the resulting fraction and peak larger. If half the families in a stage are still alive today, the denominator is half as large and the peak of percentage is twice as high as it would have been had they all been extinct. Although removing extant families affects the height of the peaks, it does not change their spacing.)

The chart clearly shows that extinction is not continuous (which geologists have known for a long time), and confirms the location of the three members of the Big Five that we know fall within the last 250 million years of earth history: the Permian–Triassic, the Triassic–Jurassic, and the K–T. Finding the three members of the Big Five exactly where they were expected to be told Raup and Sepkoski that nothing was seriously wrong with their methodology. The truly startling point, however, and the one that sent Raup rushing into Sepkoski's office, is that the peaks show up at regular intervals—every 26 million years. What could it mean?

Before answering that kind of question, when faced with such an unexpected and unprecedented pattern emerging from a complex set of data, a scientist has to make certain that the result is not merely an accident or an artifact of the way the chart was constructed. Raup spent the next several months testing these possibilities, trying to "kill the periodicity," as he put it. But no matter what he tried, the periodicity persisted, and at a confidence level of better than 99.5 percent.

Like many scientific suggestions, the idea of periodic extinction was not new. It had been proposed in 1977 by Alfred Fischer, then at Princeton, and his graduate student Michael Arthur.[6] They assembled data from a variety of such geologic indicators as sealevel, temperature, number of species through time, and isotopic ratios. Their analysis revealed a cyclical pattern in species diversity with a period of 32 million years. Fischer and Arthur did not have enough data for rigorous statistical testing, and partly for that reason their suggestion was not followed up. Since no one knew why the earth should have behaved cyclically, the observation itself was discounted.

Raup and Sepkoski, having submitted their huge volume of data to careful statistical analysis and been unable to falsify their conclusion, were ready to go public. They did so with trepidation. Raup was a distinguished, if somewhat iconoclastic, paleontologist; Sepkoski's career was barely underway. Neither wanted to become a laughingstock, or perhaps worse, to be ignored. They stuck their toes in receptive water when Sepkoski presented their preliminary findings at a 1983 symposium in Flagstaff, Arizona, home of the Astrogeology Branch of the U.S. Geological Survey (founded by Shoemaker). This friendly audience, assembled to explore the implications of the Alvarez theory, was delighted, and Sepkoski was emboldened to suggest that the source of the periodicity might be extraterrestrial. As Raup tells it, this proposal arose merely because it is much easier to find cycles in the motions of the planets, stars, and galaxies, which wheel and circle each other periodically, than to find them in apparently random earthly processes. The astronomers and astrophysicists in attendance at the meeting, intrigued by the Raup and Sepkoski analysis, set to work with a vengeance to find the cause of the 26-million-year cycle.

With no reason to delay publishing, Raup and Sepkoski chose the *Proceedings of the National Academy of Sciences (PNAS)*, the journal of that elite group of elected, eminent scientists, of which Luis Alvarez and Raup were members. The *PNAS* publishes only papers written by members, which it does not find necessary to subject to peer review. Their paper appeared in February 1984.[7]

The reaction came almost too quickly to be true. In the April 19, 1984, issue of *Nature* no fewer than five articles appeared based on the *PNAS* paper.[8] Now, although *Nature* is one of the speedier journals to publish, the submission dates on the five papers showed that they were submitted even *before* the Raup and Sepkoski paper appeared! The explanation is that Raup and Sepkoski, like most scientists, sent preprints of their submitted paper to colleagues, giving them advance warning.

One of the five papers, by Michael Rampino and Richard Stothers, confirmed the periodicity of the fossil record. Using a different statistical technique, they reanalyzed the Raup-Sepkoski data set and came up with a period of 30 ± 1 million years, which they attributed to the passage of our solar system through the plane of the Galaxy. As everyone knows, our solar system is part of the Milky Way, a vast, rotating complex of stars shaped like a planar disk—broad and spiraling when viewed from "above" but flat when seen edge on. As the Galaxy rotates, the Sun and planets move slowly up and down across the plane of the disk, the round trip taking just over 60 million years. The solar system thus crosses the galactic plane twice in each such circuit—once every 30 million years or so, not too far off the period that Raup and Sepkoski, and Rampino and

Stothers, had found. Although no one knows exactly what effect the crossing of the galactic plane has, it could be that astronomical or climatic changes are somehow induced, which, in turn, drive the periodicity. However, the Sun is now close to the galactic plane, so that there should have been a recent mass extinction, yet the latest one that Raup and Sepkoski recognized occurred in the middle Miocene, about 10 million to 11 million years ago.

The second explanation, proposed in two of the papers in *Nature,* is that the Sun has a small companion star. Because most stars that we can observe are binary, the existence of a companion would not be a surprise. The Sun's fellow traveler might be on a highly eccentric orbit that takes it far out in space but periodically brings it back nearer the outer boundaries of the solar system, where lies the Oort cloud, a vast conglomeration of comets. Although no one has seen this cloud, there is good reason for believing that it exists and that it is the source of Halley's Comet and the other "long-period" comets that approach the Sun from all over the solar system. As the putative companion star passes near the Oort cloud, its gravity could pull comets out of their present orbits and launch a few on a collision course with Earth, where they would strike, producing craters and mass extinctions. The astronomers who wrote in *Nature* agreed that the companion star must be quite small and now be located about two light years from the Sun.

But why, since the buddy star would be closer to the earth by half than any other, have astronomers never seen it? It turns out that it could easily have been missed—only a small number of stars have ever been observed and catalogued—or it might have been mistaken for a brighter star much farther away. But would its orbit have remained stable over the 250 million years of geologic history that Raup and Sepkoski examined, or would it not have been degraded by the gravity of nearby stars? One calculation showed that it could have remained constant for as much as a billion years, more than enough time.

The authors of one of the papers suggested that the companion star be named Nemesis, after the Greek goddess who punished earthly beings for attempting to usurp the privileges of the Gods.[9] In fact, they proposed other names, but the editors of *Nature* chose Nemesis and it stuck. (Muller and his co-authors noted that if the companion were never found, the paper claiming that it existed might turn out to be *their* nemesis.) Muller launched a program to search the heavens for Nemesis, using an automated telescope system that examines about 10 stars per night. So far, over 3,000 candidates have been studied, but none has yet turned out to have the characteristics of Nemesis, leaving Muller with an absence of evidence and a huge backlog of stars to go.

Stephen Jay Gould did not care for the name Nemesis and took Muller and his colleagues to task in an open letter in *Natural History*: "Nemesis is the personification of righteous anger. She attacks the vain or the powerful, and she works for definite cause. . . . She represents everything that our new view of mass extinction is struggling to replace—predictable, deterministic causes afflicting those who deserve it."[10] He proposed the star be named Siva, after the Hindu god of destruction, who, "Unlike Nemesis, . . . does not attack specific targets for cause or for punishment. Instead, his placid face records the absolute tranquillity and serenity of a neutral process, directed toward no one."[11] Siva's modus operandi comported better with the view that Gould, Raup, and others were developing in response to the Alvarez theory: Survival or extinction are essentially matters of chance, of bad luck rather than bad genes. A debate among serious scientists over which mythological name to give to a star that has never been seen and whose existence is barely even an educated guess, is one more curiosity stemming from the Alvarez theory. But perhaps it is salutary: Seldom before have paleontologists and astronomers had anything even to disagree about.

A third theory, proposed by Daniel Whitmire and Albert Jackson of the University of Southwest Louisiana, appeared soon after.[12] They suggested that the periodicity could be due to an undiscovered tenth planet, Planet X, located beyond the orbit of Pluto. Regular changes in the orbit of Planet X, about every 28 million years, could have disturbed a cloud of comets beyond the orbit of Jupiter (not the Oort cloud, which is much further out). The idea that there might be a yet undetected planet was not completely ad hoc; it had come up before as a way to explain the tiny discrepancies that remain between the calculated and observed orbits of certain planets. On the other hand, calculations show that such a planet, unless it were well outside the plane of orbit of the others, would probably be bright enough to have been detected.

Astronomers may have taken the Raup-Sepkoski periodicity to heart, but others did not. Soon after their initial paper appeared, contrary views began to arrive. This was not surprising, for as scientists know better than most, if the only way to prove your point is by using statistics, you are in trouble—especially if you are not a statistician. To live by statistics is to run the risk of dying by statistics. As Disraeli said, "There are three kinds of lies: lies, damn lies, and statistics."

Antoni Hoffman, a paleontologist at Columbia University, wrote the contrary article[13] that drew the most attention, even the blessing of John Maddox, the editor of *Nature*. Hoffman proffered three objections. First, he criticized Raup and Sepkoski for removing from their analysis species that were still living and those whose range is poorly known. This criticism is questionable, however, because although it is easy to imagine how the removal of some families could degrade an existing cyclic pattern, it is hard to see how the removal could create a strong periodicity where none existed. Surely it would merely produce more "noise." Second, Hoffman noted that because the periodicity is degraded, or disappears altogether, when a different time scale than the one used by Raup and Sepkoski is employed, their conclusion must be wrong. But this criticism is tantamount to claiming that, using an incorrect time scale, one can generate a false pattern that is periodic at a high confidence level, which seems contrary to logic. It is more likely that the degrading of the periodicity when a different scale is used means that (1) the fossil record is periodic, and (2) the time scale used by Raup and Sepkoski is closer to the true scale.

Hoffman's third argument was in a different class and purported to be the knockout punch to the proposal of periodic extinctions, and by extension, to the general notion of extraterrestrial impacts. Raup and Sepkoski had to distinguish mass extinction from the normal background extinction rate. They defined a mass extinction as having occurred whenever their data showed a rise in extinction rate from one geologic stage to the next, followed by a decline in rate in the third stage. Thus a mass extinction has occurred only when the rate of extinction is greater in a given geologic stage than in the stages above and below it. Hoffman pointed out that there is a 25 percent probability of this happening by chance. To understand his argument, label the three successive stages 1, 2, and 3. At random, stage 2 has a 50 percent probability of having a higher extinction rate than stage 1 and a 50 percent probability of having a lower rate. Stage 3 likewise has a 50 percent probability of having a higher rate than stage 2 and a 50 percent probability of having a lower one. Since probabilities multiply, $0.50 \times 0.50 = 0.25$ and the chance of stage 2 having a higher rate than either stages 1 or 3 is one in four, or 25 percent. Hoffman then went on to his clincher: 39 stages in 250 million years works out to an average stage length of 6.4 million years. But four times 6.4, rounded up a little, equals 26 million years—the periodicity found by Raup and Sepkoski! In other words, in a random set of extinction events, the 26-million-year frequency would show up 25 percent of the time, *on the average*.

This seemingly irresistible argument tempted John Maddox further out on a limb than journal editors ought to go. Not content to let the Hoffman paper speak for itself, he took the unusual step of writing an editorial comment: "The analysis is certain to yield the conclusion that, on the average, extinction peaks occur every four stages. . . . Hoffman has undermined the assumption on which all the excitement was based, the belief that there is a 26 million year periodicity to be explained." Maddox continued, "Human nature being what it is, it seems unlikely that the enthusiasts for catastrophism will now abandon their quest."[14]

The trouble with Hoffman's third argument (and Maddox's endorsement) is that they miss the point, as Stephen Jay Gould has pointed out (and from whose article the rest of the discussion in this section is drawn[15]). Hoffman wrote: "There is 0.25 probability that any particular stage represents a peak of extinction. Peaks are, then, expected to occur approximately every fourth stage."[16] Read that quotation carefully and think about what Raup and Sepkoski actually found. Hoffman is claiming that chance will produce peaks on the average every 26 million years, approximately every fourth stage. This is like saying that an honest coin, if tossed often enough, will produce heads on the average 50 percent of the time. But Raup and Sepkoski did not claim to have found a cycle with an *average* of 26 million years; they claimed to have found a peak *every* 26 million years, like a coin that, although it shows heads half the time, gives this precise sequence: HTHTHTHTHTHTHTHTHTHT. . . . Thus Hoffman's point is irrelevant to the arguments of Raup and Sepkoski, who in any case had thoroughly tested the possibility that their pattern was due to chance and rejected it at a very high confidence level. Hoffman also manipulated Sepkoski's family data using different time scales and extinction metrics, and came up with 20 different ways of gauging periodicity, on the basis of which he claimed to have falsified Raup and Sepkoski's theory. When Sepkoski subsequently combined all 20 of Hoffman's metrics, however, the 26-million-year periodicity reappeared, more robust than ever! Like the newspaper account of the death of a very-much-alive Mark Twain, the Hoffman-Maddox pronouncement of the demise of extinction periodicity was an exaggeration.

Is Cratering Periodic?

Though periodicity in the fossil record is still being criticized, Raup and Sepkoski have responded well to their critics, and as Sepkoski has added more data to his compendium, evidence for periodicity has grown stronger.[17] At the very least we can say that extinction periodicity has not been falsified. We have also seen that in theory, impact could have caused all extinction. Both ideas are far from corroborated but at least deserve the status of working hypotheses. If both are correct, impact cratering ought also to be periodic, at least in part, and on the same time cycle as the mass extinctions. In the April 19, 1984, issue of *Nature*, Walter Alvarez and astronomer Richard Muller reported that they had found a periodicity of 28 million years for terrestrial craters,[18] the same within its error as the 26-million-year cycle that Raup and Sepkoski had reported for mass extinctions. Alvarez and Muller used Grieve's 1982 compilation of terrestrial craters, selecting only those that are older than 5 million years and whose ages are known to better than ± 20 million years. Unfortunately, this tight filter produced only 13 imperfectly dated craters, too small a sample to allow their statistical conclusions to be convincing.

Reporting in the same issue of *Nature*, Rampino and Stothers applied a different statistical technique to Grieve's database and found a periodicity of 31 million years for impact craters.[19] Stothers later culled, from Grieve's list, a set of seven Cenozoic craters with age errors of less than 1 million years.[20] He compared each of the resulting ages with the ages of seven geologic stage boundaries. These data are plotted in Figure 26. (Stothers used the Manson, Iowa, crater

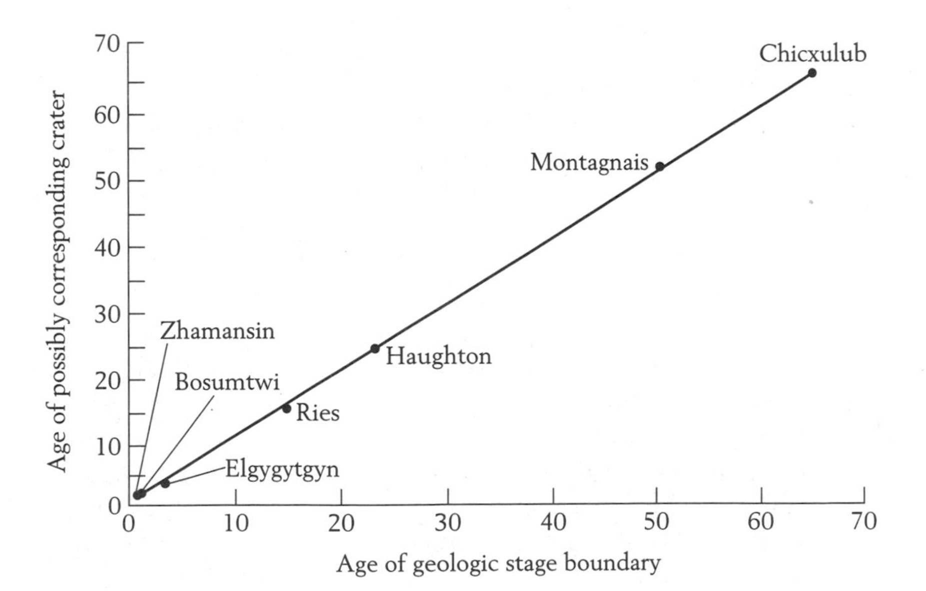

FIGURE 26 Cenozoic crater ages and geologic stages. [From data of Stothers.]

but I have substituted Chicxulub.) This chart tacitly assumes cause and effect, which may be incorrect. It really asks this question: Does the age of each of the well-dated Cenozoic impact craters match that of a geologic stage boundary? Stothers concluded that the answer is yes, and at a confidence level of 98 percent to 99 percent.

A fruitful line of research to confirm his conclusion would be to select a significant number of craters whose ages appear to lie close to the 28-million- to 32-million-year periodicity, but where the age measurement errors are too large for certainty, and to launch an intensive dating program so as to determine the ages of those craters more precisely. Grieve estimates that in order to conduct a fair test, the age uncertainties would have to come down to no more than 10 percent of the age itself. In other words, for a 30-million-year-old crater, the uncertainty would have to be no worse than ± 3 million years, well within the reach of today's technology.[21]

Other Cycles

Rampino and Stothers have gone on to argue for a 32 ± 3 million-year periodicity not only in mass extinctions and impact cratering, but in a variety of other major geologic processes: flood basalt eruptions, magnetic reversals, appearance of oxygen-poor oceans, large changes in sea level, and episodes of seafloor spreading. Later, Rampino and Bruce Haggerty went on to develop what they call their Shiva (Siva) hypothesis.[22] If they are correct, a single cause is likely to drive most or all of the earth's large-scale processes. Dare I say it? If Rampino and colleagues are right, as shown in Figure 27, they are on the trail of a grand unified theory of earth systems!

They imagine the cycle starting with the impact of an asteroid or comet, say the one that forms the 50-million-year crater, which has a diameter of 100 km. The asteroid that produced it would penetrate at least 20 km into the earth. The nearly instantaneous evacuation of a large section of the crust would relieve the pressure on the underlying mantle, causing it to melt and giving rise to floods of basalt, which would then erupt at the surface, possibly hiding the parent crater. The shock of impact would upset the magnetic dynamo in the earth's core, causing it to reverse. The ocean floor would rift and spread; sea level would fall. The poisonous effects of the gases emitted during flood basalt eruptions, added to those of impact, would cause a mass extinction and thus neatly tie the whole package together.

This takes us far out on the slenderest branch yet. One can probably count on one's digits the number of geologists who believe

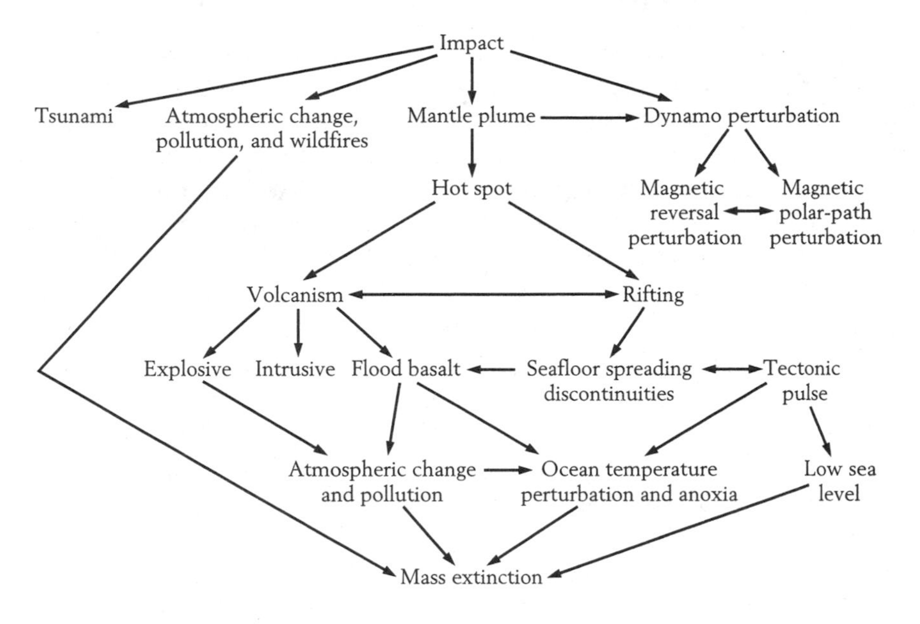

FIGURE 27 A grand unified theory of earth systems? [After Stothers and Rampino.[23]]

Rampino and Stothers are onto something. Yet recall that the K–T boundary is not the only one to have a flood basalt of nearly, if not exactly, the same age: so does the Permian–Triassic. Its age matches closely, some say identically, that of the Siberian traps. Rampino, Stothers, and most recently, French geologist Vincent Courtillot, have searched the literature for other examples of geologic boundaries with flood basalts of similar age, and have found many.[24] On the other hand, just as some craters appear to have no associated extinction, neither do some flood basalt eruptions. The boundary that we know best, the K–T, formed 1 million or 2 million years after the Deccan flood basalt volcanism began. The age of the Permian–Triassic boundary is not as well dated as the K–T, and it remains to be seen whether the claimed correspondence between the age of the Permian–Triassic and that of the Siberian traps will hold up.

PERIODICITY ASSESSED

Writing in 1989 and basing his remarks on his latest compilation of the extinction rate for genera, Sepkoski concluded that 9 of 11 extinction peaks lie on or close to the 26-million-year periodicity. The probability of this happening by chance is less than one in a million.[25] The periodicity of cratering is far less firm, due to the much smaller sample size, but it has not been falsified. Yet in spite of the evidence for extinction periodicity, and its importance if true, interest seems to have waned. One of the reasons is that periodicity falls deep in the cracks between disciplines—it is not really the province of geologists, or paleontologists, or astronomers, or statisticians, or anyone—it is a scientific orphan in a world of limited time, scarce resources, and orthodoxy. Even the pro-impactors need not endorse it—impact could be of great importance in earth history and not be periodic.

But there is clearly another reason why periodicity has failed to continue to excite scientists: No one has been able to find Nemesis, or Planet X, or any logical reason as to why extinction, and possibly cratering, should be periodic. We know from the history of science that we should not reject what seem to be sound observations just because we cannot account for them, yet it is human nature to do so. Recently, however, an intriguing explanation has come to light that may mean that a plausible source mechanism is available.

Astronomers have been studying the effect of passage of the Oort cloud—that vast reservoir of comets out beyond the solar system—through the disk of the Galaxy, and have found previously unrecognized gravitational effects that could cause comets to strike periodically and that would repeat every 30 million to 35 million years.[26,27] Previously, Shoemaker was on record as thinking that cratering periodicity was a statistical fluke, but this new work made a believer out of him: "Impact surges are real . . . and [the comet flux is] controlled by the fluctuating galactic tides," he said. The new work "is a landmark contribution in understanding the history of bombardment of the earth."[28] Add this to the opinion of paleontologist Douglas Erwin that "The periodic signal continues to shine through the turmoil, battered but resilient," and we can see that periodicity of both the fossil record and terrestrial cratering are hypotheses that are alive and well.[29]

Chapter 13

Geology's Golden Age

The impact of solid bodies is the most fundamental of all processes that have taken place on the terrestrial planets.[1]
Eugene Shoemaker

A Second Revolution?

In the last three decades, geologists have been asked to accept, in order, that continents are not fixed in place but, carried on giant plates, roam over the surface of the earth; that impact is ubiquitous in our solar system; that thousands of meteorites, some of them huge, have struck the earth in its history; and that one impact formed the Chicxulub crater and caused the K–T mass extinction. If the fossil record is periodic, which the evidence strongly suggests it is, geologists will also be asked to consider the likelihood that several mass extinctions, and not just one, are due to extraterrestrial impact. Thus to a greater extent than even the pro-impactors could have imagined when the Alvarez theory broke, there is strong evidence that major events in earth history are controlled by forces from outside the earth. Where do these advances leave what has been the key concept of geology for a century and a half: uniformitarianism?

Recall from Chapter 2 that the awkward term *uniformitarianism* was coined by Whewell to describe Lyell's conception of the earth. Lyell believed that the only processes that have ever operated are those that we can observe operating today, which have always operated at the same rate. As a result, the earth has always looked as it does now; its history reveals no evidence of directional change. His uniformitarianism of rate and state was disproven more than a century ago and abandoned, but geologists apotheosized Lyell's uniformitarianism of process and natural law in the words of James

Hutton: "The present is the key to the past." Catastrophism was rejected; Lyell and subsequent geologists needed no "help from a comet."

Of course, the slow processes that we see today at the earth's surface—wind- and water-driven erosion and deposition, the advance and retreat of glaciers and the sea—can be projected back into the past, and, in that sense, the present is at least a part of the key to the past. Thus a case could be made for retaining the uniformitarianism of process. But the whole edifice has caused such damage, and is today so misleading, that the case for abandoning uniformitarianism is much the stronger. Strict adherence to uniformitarianism clearly played a role in delaying for half a century the recognition of continental drift and its modern version, plate tectonics. As for meteorite impact, Marvin writes, that "Uniformitarianism . . . probably has been the single most effective factor in preventing geologists from accepting the idea . . . as a process of any importance in the evolution of the earth."[2] If we measure from the date of Gilbert's erroneous conclusion about the origin of Meteor Crater in 1891, to 1980, the year of the first Alvarez paper, uniformitarianism and anticatastrophism cost geology nearly 90 years. But continental drift and meteorite impact are arguably the two most important processes that have affected the history of the earth. How much value remains in a paradigm that helped to retard the recognition of both for generations?

All scientists, geologists included, study cause and effect and then project to cases where only effect can be seen. But that is all that the uniformitarianism of process amounts to, and it is drastically incomplete. Is there any longer a reason for geologists, alone among all scientists, to give an exalted name to the standard modus operandi of science? Doing so is more apt to misdirect geologists of the future, and to load them with the baggage of the past, than to assist them in understanding the history of our planet.

Where do the traditional uniformitarian explanations of mass extinction—changes in climate and sea level—stand today? Has additional evidence been uncovered since the Alvarez theory appeared in 1980 that lends them greater credence? No, just the opposite. While the Alvarez theory has grown stronger, they have grown weaker. I noted earlier that David Jablonski found that major mass extinctions failed to correlate in any way with known changes in sea level, global climate, and mountain building.[3] Recently, John Alroy of the Smithsonian compared the appearance and extinction of mammals with the ups and downs of global climate over the past

80 million years.[4] He found that until 65 million years ago, mammal diversity was low, and that in the aftermath of the Chicxulub impact it fell even lower. The number of species then rose sharply to reach a plateau about 50 million years ago, where it has remained since. Almost no correlation exists, however, between climate and the appearance and extinction of mammals. Instead, the appearance of new species was largely controlled by the number already present: When mammals were few to start with, more new species appeared. The driver of mammalian diversity thus seems to be not climate but the number of vacant ecological niches. Alroy's study does not rule out the possibility that some mass extinctions have been caused by rare extremes of climate. But taking the view that it is what a species does know that cannot hurt it, Alroy noted that cyclical changes in the position and shape of the earth's orbit relative to the sun produce changes in climate every 20,000, 40,000, and 100,000 years. (Most geologists believe these are the causes of the repeating ice ages.) Species, which live on the average for a few million years, have of necessity survived scores of changes in climate and sea level.

In summary, it seems fair to say that, nearly two centuries after Hutton, there is precious little positive evidence that changes in climate and sea level cause mass extinctions. It is up to the proponents of the claim that they produce such evidence.

CHANCE IN EARTH HISTORY AND IN SCIENTIFIC DISCOVERY

Not only have the advances since the Alvarez theory appeared brought about a transformation of geology, they have greatly illuminated the role of chance in our solar system. We see that not only the death of the dinosaurs, but our presence on the earth, is contingent on the particular way in which the solar system originated and evolved. The K–T impact was set in motion nearly 4.5 billion years ago with the birth of the solar system. From that primordial chaos arose a comet or an asteroid that through the subsequent eons was intermittently pounded by impact and continually nudged by gravity. Had one collision been just a bit more or a bit less energetic, had gravity tugged a little more here or a little less there, the impactor would have had a different size and a different orbit. The dinosaur killer would have struck at some other time in the earth's history, or missed our planet entirely, and dinosaurs would not have become extinct when they did. Who knows, perhaps their 160-million-year reign might have stretched to 225 million years—and they would still be alive today. If 65 million years ago the mammals had found a no vacancy sign, I would not be writing and you would not be reading—our species would not exist.

This view of life in the solar system suggests that evolution can be more a matter of chance than inevitability. The dinosaurs did not expire because of a fatal flaw while the flawless mammals lived on. Dinosaurian genes were not inferior to mammalian ones. Life after the K–T event was not an improvement on life before and did not necessarily represent Progress with a capital P. It may be instead that, after the fall, our small, furtive ancestors survived by skulking in burrows and crevices and eating the remains of other creatures, many of whom might have seemed superior to them.

Even the discovery of the Alvarez theory might itself have been due largely to chance. Whether we think it was depends on how we see the work of Jan Smit. The discovery of iridium in the Gubbio boundary clay by the Alvarezes was serendipitous to be sure, but Smit was on the right trail. Had the Alvarezes not gotten there first, would the high iridium levels in his Caravaca samples, hidden in the archives at Delft, ever have come to light, and led him to propose the Smit theory? That we shall never know.

YOUNG TURKS AND OUTSIDERS

Thomas Kuhn said that those "who achieve . . . fundamental inventions of a new paradigm have been either very young or very new to the field whose paradigm they change."[5] The former are often known as young turks, the latter as outsiders. The history of science is full of examples of the vital role both play. Take geology: Alfred Wegener was an outsider—a meteorologist and polar explorer—who conceived the idea of continents floating through the mantle as he watched icebergs drift across the arctic seas. Although continents do not float in the mantle, in the largest sense he was right—they do move—but it took half a century for insiders to wake up to it. Luis Alvarez was a physicist without whose intrusion we might still be saying (if not really believing) that sea level changes killed off the resilient dinosaurs.

Outsiders can provide an indispensable point of view. With few exceptions, scientists work within the current paradigm, which permeates construction of their theories, even their approach to thinking about their subject. They know which old questions need not be asked again, and fail to see which new ones might fruitfully be raised. Their philosophy may prevent them from taking even the first small step on the journey to a paradigm shift. But newcomers—either young in age or new to the field—are unburdened by the weight of the prevailing paradigm. Indeed, the outsider often does not know enough to work within the paradigm even if he or she wanted to. Typically the outsider has neither the background nor the interest required to learn a new field from scratch. Why climb the mountain in one field, as Luis did, only to descend so that you can laboriously pack the gear of climbers who are scaling a new mountain? Better to leap from peak to peak.

Much of the interesting work in science, as the Alvarez theory shows so well, is done at the interface between disciplines. Progress is made when the techniques of one discipline are applied for the first time, or in novel ways, to the problems of another discipline, something that outsiders are in a good position to do. Nobelist Harold Urey brought his expertise as a chemist to bear on problems of the earth sciences and made many outstanding contributions. Outsiders are like bees carrying vital scientific pollen from one disciplinary flower to another.

Another facet of the case for the outsider is that the young are apt to be overly influenced by the stifling presence of the magister. Following the lead of their elders and their own self-interest, young scientists naturally pursue what they see as possible, which by definition usually lies within the current paradigm. But outsiders, particularly Nobelists such as Urey and Alvarez, whose respectability is not in question and who owe allegiance neither to the magisters of the field nor to the ruling paradigm, can step in with impunity. Indeed, it seems likely that nothing would have made Luis Alvarez happier than to break rank with Lyell and Hutton; had he been a 30-year-old geologist in his first position, however, it might have been different.

Seldom does a magister launch a paradigm shift within the field in which his or her eminence was achieved. To do so would mean casting off previous work and conclusions—tantamount to admitting error or poor judgment. But a magister who turns out to have been wrong may no longer deserve the title. Few have been able to walk the tightrope of maintaining eminence while correcting past errors of judgment.

The Power of Dissent

Francis Bacon captured a key aspect of science when he said that "Truth emerges more readily from error than from confusion."

Science learns from its mistakes. To find them, scientists must criticize, or dissent, at least for a while. Outsiders, not being caught up in the mores and personal relationships of their newly chosen discipline, are in a particularly strong position to dissent from the prevailing view. The best scientists dissent from even their own conclusions, as when Luis invented a new theory every week and (successfully for a while) shot each down in turn, or when Raup tried to "kill the periodicity." Only after they have been unable to falsify their own results do they publish. When scientists initially fail to dissent from their own still tentative conclusions (often by avoiding the obvious, definitive test), they run the risk of dishonoring themselves and forsaking their discipline. The false claims of cold fusion provide the clearest recent example.

Styles of dissent run the gamut from friendly critic to bitter enemy. Although personal relations may suffer, science ultimately cares little about the form and style of dissent as long as some general rules are followed. Nice people and nasty ones alike can finish first, last, or in the middle. Among the rules are these: Criticism is to be based on new evidence or on a better interpretation of the old evidence. Rebuttals are not only to be voiced at professional meetings, they are to be written up and submitted for peer review and publication. *Ad hominem* attacks are frowned upon. Ideally, opponents share data, microscopes, and outcrops. Blind tests are cheerfully conducted. And so on.

This brings us naturally to the role of Charles Officer, that most vociferous and untiring critic of the Alvarez theory. His opposition culminated in 1996 with publication of his book with Jake Page, *The Great Dinosaur Extinction Controversy.*[6] His dogged, constant, and long lasting resistance is bound to tell us something about how science works.

How far, for instance, will a scientist on the losing end of an argument go? Judging from his book with Page, Officer is willing to go so far as to leave science altogether. Officer's and Page's overall position is given away by this astounding statement: "Most of the 'science' performed by the Alvarez camp has been so inexplicably weak, and the response to it so eagerly accepting by important segments of the scientific press, never mind the popular press and the tabloids, that some skeptics have wondered if the entire affair was not, on the impact side, some kind of scam."[7] They go on to employ a set of stratagems that seem hauntingly familiar; suddenly one realizes that they are the very ploys used by creationists and others who have no platform of logic. They try, for example, the Confident Assertion: "One of the things that did not happen at the K–T boundary was an impact by a gigantic meteorite,"[8] and The Strawman: "There was no big dinosaur bone pile . . . that might have resulted from an instantaneous event."[9] (Scientists have shown that the K–T extinction would not have produced large bone piles.) They resort to the Red Herring: There "is a connection between livestock problems and the demise of the dinosaurs,"[10] and plead for equal time: "Between 1991 and 1993 . . . *Science* published eleven articles favorable to [impact] and two unfavorable."[11] They blame the media: "Before long the bias [of *Science*] was so evident to members of the Earth science community that few even bothered to submit . . . a manuscript that espoused a terrestrial cause"[12]; and they impugn the motives of the pro-impactors: "In degenerating [research] programs . . . theories are fabricated only in order to accommodate known facts." They conclude that the Alvarez theory is "not merely pathological science but dangerous to boot."[13]

In courtrooms, legislative halls, and debating tournaments, the more determined and skillful an argument on one side, the more the position of the other side is weakened. Even in the face of a mountain of evidence, an adroit defense attorney can see a guilty man set free. It would be reasonable to assume that Officer's long struggle has weakened the Alvarez theory and that, one day, Officer may overthrow Alvarez. But here we find another way in which science differs: Far from weakening the Alvarez theory, Officer's dissent has greatly strengthened it. Officer's papers were accepted and published in respectable journals, requiring the pro-impactors to polish up their thinking and respond. As a result, we now know far more about the geochemistry of iridium than if Officer and others had accepted from the start that it is indeed a marker of impact not found in volcanic rocks. We now know much more firmly that multiple sets of planar deformation features are caused only by impact. Blind tests have been conducted that otherwise would have been deemed a waste of time.

With an irony worthy of Greek tragedy, Officer's tireless, obsessive battle has had just the opposite outcome than he intended; its main effect has been to cause doubters to reserve judgment and to wait for stronger evidence to support impact, which eventually came. Today, hardly anyone other than Officer doubts the existence of the Chicxulub crater, though, as noted, some paleontologists do doubt that it is linked to the K–T mass extinction. Officer's role is different from that of, say, G. K. Gilbert, or from the authorities who opposed continental drift from the 1920s through the 1960s. When those magisters pronounced that terrestrial craters were caused by gas explosions from below, or that continents cannot move, research was shut down for half a century or more. Officer's opposition, and especially his style, made the pro-impactors try all the harder.

WHERE FROM HERE?

Science and evolution both operate as punctuated equilibria. Almost all scientists work to extend and perfect the prevailing paradigm, and continue doing so until a new discovery, often made by accident, requires that the paradigm be reexamined. At first, attempts are made to fit the new discovery in, and often they succeed for a while. But gradually it becomes clear to the more progressive practitioners of a discipline that the old paradigm simply cannot explain enough of the new evidence and must be replaced. The progress of science is then punctuated by the arrival of a new paradigm, which in most cases was developing offline, like a shadow government, ready to step in when needed.

Just after the arrival of a new paradigm, things are muddled and confused. Some questions have been answered but more have been raised. Like species after the punctuation of biological equilibrium, science is now evolving rapidly. It is not always a pretty sight as some continue to hang back while others shoulder in. Because it is hard to know which research directions are apt to be the most fruitful, false leads are followed and dead-end sidings are entered. The old methods and theories prove unable to explicate the new paradigm and new methods have to be invented. The immediate aftermath of the arrival of a new paradigm presents many niches of opportunity into which thc nimble, the young turks, and the outsiders can move. (This was the state of physics during the 1930s and 1940s, of which a young turk named Luis Alvarez took full advantage.) In time, these birthing pangs recede, scientists turn to extending and perfecting the new paradigm, and the cycle begins anew.

Earth scientists know that this is the way a paradigm shifts, for between 1966 when plate tectonics arrived and, say, 1976, when it was fully developed and accepted, we were witnesses. Now, with the Alvarez theory just a decade-and-a-half old and the crater discovered only in the 1990s, geology once again finds itself in a time of great opportunity. If impact has played the broader role hinted at in these chapters, important discoveries may lie just ahead.

What do we know today about the role of meteorite impact? We know that it was the dominant process in the primordial solar system. As the objects that had just condensed from the solar dust cloud collided with each other, sometimes fragmenting and sometimes adhering, the inner planets were born. For hundreds of millions of years thereafter, impact continued alternately to destroy and to rebuild their surfaces. One giant collision even carved the moon from the earth. The early bombardment was so intense that the surfaces of the inner planets and their satellites melted completely. Nothing escaped the inevitability of impact. Those objects that appear at first glance to have avoided it, for example, certain of Jupiter's moons, turn out to have had recent volcanic activity or to be covered with ice, obscuring the underlying craters. Every object in the solar system has been shaped by myriad collisions. Three decades of research have proven Gene Shoemaker right: Impact is "the most fundamental process."

The impact of comets and asteroids on the earth might not only have destroyed life, it might have delivered it. The K–T boundary clay contains amino acids not found elsewhere on our planet; perhaps the early impacting comets brought with them other building blocks of life that then combined and evolved to colonize Earth. Or, perhaps life developed first on Mars and was brought to Earth by a chunk of rock blasted off the red planet by impact. These are among the exciting possibilities that scientists will be studying over the next few years.

If impact is fundamental in the solar system taken as a whole, Earth could not have escaped. We have discovered about 160 impact craters, of which one—Chicxulub—was formed in the most energetic event in the last billion years of earth history. Even though it is counterintuitive, our intellect requires that we recognize that Earth has been struck many more times than 160; it must have been hit thousands, indeed tens of thousands of times. But where is the evidence of these collisions and their effect on Earth and on life? Could 50 million bombs the size of the one dropped on Hiroshima exploding every 7 million years, and larger events less often, have had no effect on life? So far, the evidence is insufficient to answer this question. This contradiction between reason and observation could have one of three explanations, or, more likely, a combination of all of them: First, most of the evidence of impact may have been removed by erosion; second, our methods of detecting impact may be inadequate; or third, we may not have looked systematically enough. This third possibility represents an opportunity.

So far in this story, advances have come about in the traditional way: through the efforts of scientists working alone or in small groups, each following their intellectual curiosity, without an overall strategy. Though many scientists would agree with Al Fischer, who does not like "science by committee," one can still ask whether this style of research is most apt to produce rapid progress in exploring the implications of the Alvarez theory. Given the disparate interests of scientists, and what we now know to be the complexity of the questions, the difficulty of reading the geologic record, and the scarcity of funding, there is a case for focusing resources and proceeding strategically. One idea would be to establish a Center for the Study of Impact and Extinction where scientists from a variety of disciplines could come together. The National Science Foundation funds such centers in other fields on university campuses; models exist and they have proven effective.

What would such a center do? Two lines of research are essential. The greatest obstacle to progress is that the ages of geologic boundaries, extinction horizons, impact craters, and flood basalts are not known with sufficient precision or accuracy to permit firm conclusions. The first need, then, is to improve techniques of age measurement. The argon-argon method is the most precise (most reproducible), but its accuracy (closeness to the true value) can be improved.

Rather than a large number of boundaries and possibly corresponding craters being studied more superficially, a selected few should be dated and explored in depth. Horizons that appear to have corresponding flood basalts should be chosen and the ages of both the basalts and the boundary pinned down precisely and accurately. Perhaps the most immediate payoff would come from precisely dating several impact craters that now appear to be the result of periodic impacts, but where poor age precision leaves an uncertainty. If it could be established with statistical rigor that, say, a dozen craters were periodic, the periodicity of extinction, though not directly proven, would be much more plausible.

The second fruitful direction, suggested by Peter Ward, would complement the first.[14] Until now, scientists have started with evidence of impact and searched for the parental crater. This is how Chicxulub was found, but how easy it would have been to miss! Working in the other direction may be more productive: Start with a few of the craters selected for precise age dating, and look for impact and extinction effects at the corresponding levels in the geologic column. Once the age of a crater is pinned down, geologists will know where to look in stratigraphic sections to find the corresponding effects. If, at the predicted level, no impact evidence is found, when geologic techniques have improved enough so that we can be reasonably certain the evidence would have been found had it existed (the flaw now), the K–T event would appear singular and impact would somehow be of lesser import in earth history than we thought. Back to the drawing board. On the other hand, if through such methods the periodicity of cratering could be corroborated, and if three or four craters could be tied to specific extinctions, the Raup-Shoemaker impact-kill curve (see Figure 24) could be roughly calibrated and at least its overall shape determined, giving graphic form to a scientific revolution.

We have seen how a young geologist in Italy, studying something else, decided to bring home for his father a specimen that captured one of the major events in earth history. Thus was launched a scientific partnership that, conjoined with the work of hundreds of proponents and opponents alike, led to the solution of a great mystery. Today we have gone about as far as science can go in corroborating the notion that the impact of a meteorite caused the extinction of the dinosaurs. But as always, answering one set of questions raises others, and we are left pondering the true role of impact. As even its bitterest opponents have to admit, the Alvarez theory has brought geology not only a new set of questions, but a greatly improved set of sampling techniques and analytical methods for answering them. Paleontologists collect much larger samples and subject them to statistical tests. Today geologists know how to find and identify terrestrial craters. These are the hallmarks of a fertile theory.

In 1996, science writer John Horgan published a highly controversial book, *The End of Science,*[15] in which he argued that such disciplines as physics, cosmology, evolutionary biology, social science, and chaos theory, have run into intellectual cul-de-sacs, are no longer productive, and therefore have come to their natural end. Whether or not one is persuaded by his argument, it is significant that Horgan mentions not a single example from the earth sciences. Far from coming to an end, beginning with plate tectonics in the 1960s, moving on to incorporate the advances of the space age, continuing today with the exploration of the Alvarez theory, and proceeding on tomorrow to determine the true place of impact and the causes of mass extinction, geology is in its golden age.

References

(Page numbers for quotations are given in parentheses.)

Preface

1. Krogh, T. E., S. Kamo, et al. (1993). "U-Pb Ages of Single Shocked Zircons Linking Distal K/T Ejecta to the Chicxulub Crater." *Nature* **366:** 731–734.

Prologue

1. Jepsen, G. L. (1964). "Riddles of the Terrible Lizards." *American Scientist* **52:** 227–246. (p. 231)
2. Leakey, M. (1996). Personal communication.
3. Dodson, P. (1990). "Counting Dinosaurs; How Many Kinds Were There?" *Proceedings of the National Academy of Sciences of the United States of America* **87:** 7608–7612.
4. Gartner, S., and J. P. McGuirk (1979). "Terminal Cretaceous Extinction Scenario for a Catastrophe." *Science* **206:** 1272–1276. (p. 1276)
5. Eliot, T. S. (1952). *The Complete Poems and Plays.* New York: Harcourt, Brace and Company. (p. 59)
6. Russell, D. A. (1979). "The Enigma of the Extinction of the Dinosaurs." *Annual Review of Earth and Planetary Sciences* **7:** 163–182.
7. Ibid. (p. 178)
8. Kuhn, T. S. (1970). *The Structure of Scientific Revolutions.* Chicago: University of Chicago Press.
9. Shoemaker, E. M. (1984). "Response to Award of Gilbert Medal." *Bulletin of the Geological Society of America* **95:** 1002.

Chapter 1

1. Pasteur, L. (1854). Address at the University of Lille, December 7, 1854.
2. Alvarez, L. W. (1987). *Adventures of a Physicist.* New York: Basic Books.
3. Medawar, S. P. (1967). *The Art of the Soluble.* London: Methuen. (p. 21)
4. Alvarez, L. W. (1987). *Adventures of a Physicist.* New York: Basic Books.
5. Ibid. (p. 8)
6. Ibid. (p. 252)

7. Alvarez, W., M. A. Arthur, et al. (1977). "Upper Cretaceous–Paleocene Magnetic Stratigraphy at Gubbio, Italy." *Bulletin of the Geological Society of America* **88:** 383–389.
8. Alvarez, L. W. (1987). *Adventures of a Physicist.* New York: Basic Books. (p. 253)
9. Ibid.
10. Alvarez, L. W., W. Alvarez, et al. (1980). "Extraterrestrial Cause for the Cretaceous–Tertiary Extinction." *Science* **208:** 1095–1108.
11. Clarke, A. C. (15 February 1971). "Putting the Prophets in Their Place." *Time.* (p. 38)
12. Ehrlich, P. R., J. Harte, et al. (1983). "Long-Term Biological Consequences of Nuclear War." *Science* **222:** 1293–1300.
13. Smit, J. (1996). Personal communication.
14. Ibid.
15. West, S. (1979). "An Iridium Clue to the Dinosaurs' Demise." *New Scientist.* **82:** 798.
16. Smit, J., and J. Hertogen (1980). "An Extraterrestrial Event at the Cretaceous–Tertiary Boundary." *Nature* **285:** 198–200. (p. 198)
17. Roberts, R. M. (1989). *Serendipity: Accidental Discoveries in Science.* New York: John Wiley.

CHAPTER 2

1. Whitehead, A. N. Attributed.
2. Alvarez, L. W. (1983). "Experimental Evidence that an Asteroid Impact Led to the Extinction of Many Species 65 Million Years Ago." *Proceedings of the National Academy of Sciences* **80:** 627–642.
3. Marvin. U. B. (1990). Impact and Its Revolutionary Implications for Geology. *Global Catastrophes in Earth History,* eds. V. L. Sharpton and P. D. Ward. Boulder, Colo.: Geological Society of America. (**Special Paper 247:** 147–154.)
4. Ibid. (p. 152)
5. Planck, M. (1959). *The New Science.* New York: Meridian Books. (p. 299)
6. Whiston, W. (1696). *A New Theory of the Earth, from Its Original to the Consummation of all Things.* London: R. Roberts.
7. Gould, S. J. (1987). *Time's Arrow, Time's Cycle.* Cambridge, Mass.: Harvard University Press. (p. 139)
8. Darwin, C. (1964 Facsimile Edition). *On the Origin of Species by Means of Natural Selection.* Cambridge, Mass.: Harvard University Press. (p. 82)
9. Colbert, E. H. (1961). *Dinosaurs: Their Discovery and Their World.* New York: Dutton. (p. 252)
10. Bucher, W. H. (1963). "Cryptoexplosion Structures: Caused from Without or from Within the Earth? ("Astroblemes" or "Geoblemes?")." *American Journal of Science* **261:** 597–649. (p. 643)

11. Hallam, A. (1981). "The End-Triassic Bivalve Extinction Event." *Palaeogeography, Palaeoclimatology, Palaeoecology* **35**:1–44. (p. 36)
12. McCartney, K., and J. Nienstedt (1986). "The Cretaceous/Tertiary Extinction Controversy Reconsidered." *Journal of Geological Education* **34**:90–94. (p. 92)
13. Editorial (April 2, 1985). "Miscasting the Dinosaur's Horoscope." *New York Times,* p. A26.
14. Gould, S. J. (1985). All the News that's Fit to Print and Some Opinions that Aren't. *Discover,* **November 1985:** 86–91. (p. 91)
15. Good, M. L. (1997). A Second American Century? *Chemical and Engineering News* **75**, April 14, 1997: 39–45. (p. 40)
16. Hutton, J. (1795). *Theory of the Earth,* Vols. 1 & 2, Edinburgh: William Creech. (Vol. II, p. 547)
17. Lyell, C. (1830–1833). *Principles of Geology, Being an Attempt to Explain the Former Changes of the Earth's Surface by Reference to Causes Now in Operation.* London: John Murray.
18. Gould, S. J. (1987). *Time's Arrow, Time's Cycle.* Cambridge, Mass.: Harvard University Press. (p. 105)
19. Marvin, U. B. (1990). Impact and Its Revolutionary Implications for Geology. *Global Catastrophes in Earth History,* eds. V. L. Sharpton and P. D. Ward. Boulder, Colo.: Geological Society of America. (**Special Paper 247:** 147–154.)
20. Lyell, C. (1830–1833). *Principles of Geology, Being an Attempt to Explain the Former Changes of the Earth's Surface by References to Causes Now in Operation.* London: John Murray. (Vol. II, pp. 164–165)
21. Ibid. (Vol. I, p. 123)
22. Gould, S. J. (1987). *Time's Arrow, Time's Cycle.* Cambridge, Mass.: Harvard University Press. (p. 119)
23. Whewell, W. (1832). "Principles of Geology . . . by Charles Lyell." *Quarterly Review* **47:** 103–132. (p. 126)
24. Ibid. (p. 126)
25. Gould, S. J. (1987). *Time's Arrow, Time's Cycle.* Cambridge, Mass.: Harvard University Press. (p. 171)
26. Gould, S. J. (1965). "Is Uniformitarianism Necessary?" *American Journal of Science* **263:** 223–228.
27. Marvin, U. B. (1990). Impact and Its Revolutionary Implications for Geology. *Global Catastrophes in Earth History,* eds. V. L. Sharpton and P. D. Ward. Boulder, Colo.: Geological Society of America. (**Special Paper 247:** 147–154). (p. 154)
28. Alvarez, W. (1986). "Toward a Theory of Impact Crises." *Eos* **67:** 649–658. (p. 654)

CHAPTER 3

1. Lyell, K. M. (1881). *Life, Letters, and Journals of Sir Charles Lyell.* London: John Murray. (pp. 261–262)

2. de Maupertuis, P. L. M. (1750). *Les Oeuvres de M. de Maupertuis.* Dresden: Librairie du Roy.
3. Laplace, P. S. de (1813). *Exposition du Système de Monde.* Paris: Mme. V_e Courcier.
4. McLaren, D. (1970). "Time, Life, and Boundaries." *Journal of Paleontology* **44:** 801–815.
5. Urey, H. C. (1973). "Cometary Collisions and Geological Periods." *Nature* **242:** 32–33.
6. Hartmann, W. K. (1985). "Giant Impact on Earth Seen as Moon's Origin." *Geotimes* **30:** 11–12.
7. Daly, R. A. (1946). "Origin of the Moon and Its Topography." *Proceedings of the American Philosophical Society* **90**:104–119.
8. Marvin, U. B. (1986). "Meteorites, the Moon, and the History of Geology." *Journal of Geological Education* **34:** 140–165. (p. 146)
9. Gilbert, K. G. (1896). "The Origin of Hypotheses. Illustrated by the Discussion of a Topographic Problem." *Science, New Series* **3:** 1–24.
10. Marvin, U. B. (1986). "Meteorites, the Moon, and the History of Geology." *Journal of Geological Education* **34:** 140–165. (p. 155)
11. Wilhelms, D. E. (1993). *To a Rocky Moon.* Tucson: University of Arizona Press.
12. Ibid. (p. 20)
13. Marvin. U. B. (1990). Impact and Its Revolutionary Implications for Geology. *Global Catastrophes in Earth History,* eds. V. L. Sharpton and P. D. Ward. Boulder, Colo.: Geological Society of America. (**Special Paper 247:** 147–154.)
14. Dietz, R. S. (1961). "Astroblemes." *Scientific American* **205:** 51–58.
15. Dietz, R. S. (1964). "Sudbury Structure as an Astrobleme." *Journal of Geology* **72:** 412–434.
16. French, B. M., and N. W. Short, Eds. (1968). *Shock Metamorphism of Natural Materials.* Baltimore: Mono Book Corporation.
17. Levy, D. H., E. M. Shoemaker, et al. (1995). "Comet Shoemaker–Levy 9 Meets Jupiter." *Scientific American* **273:** 84–91.
18. Shoemaker, E. M., R. F. Wolfe, et al. (1990). Asteroid and Comet Flux in the Neighborhood of the Earth. *Global Catastrophes in Earth History,* eds. V. L. Sharpton and P. D. Ward. Boulder, Colo.: Geological Society of America. (**Special Paper 247:** 155–170.)

CHAPTER 4

1. Popper, K. R. (1968). *The Logic of Scientific Discovery.* New York: Harper and Row. (p. 41)
2. Tennyson, Lord A. (1955). *In Memoriam.* London: Macmillan. (p. 1)
3. Popper, K. R. (1968). *The Logic of Scientific Discovery.* New York: Harper and Row.
4. Silver, L. T., and P. H. Schultz, Eds. (1982). *Geological Implications of Impacts of Large Asteroids and Comets on the Earth.* Boulder, Colo.:

Geological Society of America. Sharpton, V. L., and P. D. Ward, Eds. (1990). *Global Catastrophes in Earth History.* Boulder, Colo.: Geological Society of America. Ryder, G., D. Fastovsky, et al., Eds. (1996). *The Cretaceous–Tertiary Event and Other Catastrophes in Earth History.* Boulder, Colo.: Geological Society of America.

5. Orth, C. J., J. S. Gilmore, et al. (1981). "An Iridium Abundance Anomaly at the Palynological Cretaceous–Tertiary Boundary in Northern New Mexico." *Science* **214:** 1341–1343.
6. Kyte, F., and J. T. Wasson (1986). "Accretion Rate of Extraterrestrial Material: Iridium Deposited 33 to 67 Million Years Ago." *Science* **232:** 1225–1229.
7. Gostin, V. A., R. R. Keays, et al. (1989). "Iridium Anomaly from the Acraman Impact Ejecta Horizon: Impacts Can Produce Secondary Iridium Peaks." *Nature* **340:** 542–544.
8. Dypvik, H., S. T. Gudlaugsson, et al. (1996). "Mjolnir Structure: An Impact Crater in the Barents Sea." *Geology* **24:** 779–782.
9. Bohor, B., E. E. Foord, et al. (1984). "Mineralogic Evidence for an Impact Event at the Cretaceous–Tertiary Boundary." *Science* **224:** 867–869.
10. Olsson, R., K. G. Miller, et al. (1997). "Ejecta Layer at the Cretaceous–Tertiary Boundary, Bass River, New Jersey." *Geology* **25:** 759–762.
11. Wolbach, W., R. Lewis, et al. (1985). "Cretaceous Extinctions: Evidence for Wildfires and Search for Meteoritic Material." *Science* **230:** 167–170.
12. Carlisle, D. B. (1995). *Dinosaurs, Diamonds, and Things from Outer Space.* Stanford, Calif.: Stanford University Press.
13. Raup, D. M. (1986). *The Nemesis Affair.* New York: Norton (p. 81)
14. Luck, J. M., and K. K. Turekian (1983). "Osmium-187/Osmium-186 in Manganese Nodules and the Cretaceous–Tertiary Boundary." *Science* **222:** 613–615.
15. Peuckerehrenbrink, B., G. Ravizza, et al. (1995). "The Marine Os-187/Os-186 Record of the Past 80 Million Years." *Earth and Planetary Science Letters* **130:** 155–167.
16. Carlisle, D. B., and D. R. Braman (1991). "Nanometre-Size Diamonds in the Cretaceous/Tertiary Boundary Clay of Alberta." *Nature* **352:** 708–709. Carlisle, D. B. (1995). *Dinosaurs, Diamonds, and Things from Outer Space.* Stanford, Calif.: Stanford University Press.
17. Broad, W. J. (June 11, 1996). "Glittering Trillions of Tiny Diamonds Are Shed by the Crash of Asteroids." *New York Times,* p. B5.

CHAPTER 5

1. Safire, W., and L. Safir (1982). *Good Advice.* New York: Quadrangle Books. (p. 44)

2. Officer, C. B., and C. L. Drake (1983). "The Cretaceous–Tertiary Transition." *Science* **219:** 1383–1390. Officer, C. B., and C. L. Drake (1985). "Terminal Cretaceous Environmental Events." *Science* **227:** 1161–1167.
3. Glen, W. (1982). *The Road to Jaramillo.* Stanford, Calif.: Stanford University Press.
4. Berggren, W. A., D. V. Kent, et al. (1995). A Revised Cenozoic Geochronology and Chronostratigraphy. *Geochronology, Time Scales, and Global Stratigraphic Correlation*, eds. W. A. Berggren, D. V. Kent, M.-P. Aubry, and J. Hardenbol. Tulsa, Okla.: Society for Sedimentary Geology. **54:** 129–212.
5. Silver, L. T., and P. H. Schultz, Eds. (1982). *Geological Implications of Impacts of Large Asteroids and Comets on the Earth*. Boulder, Colo.: Geological Society of America.
6. Drake, C. (1982). "Impacts and Evolution Conference: Reports and Comments." *Geology* **10:** 126–128.
7. Officer, C. B., and C. L. Drake (1983). "The Cretaceous–Tertiary Transition." *Science* **219:** 1383–1390. (p. 1384)
8. Alvarez, W., L. W. Alvarez, et al. (1984). "The End of the Cretaceous: Sharp Boundary or Gradual Transition?" *Science* **223:** 1183–1186.
9. Ibid.
10. Officer, C. B., and C. L. Drake (1983). "The Cretaceous–Tertiary Transition." *Science* **219:** 1383–1390. (p. 1388)
11. Archibald, J. D., R. F. Butler, et al. (1982). "Upper Cretaceous–Paleocene Biostratigraphy and Magnetostratigraphy, Hell Creek and Tullock Formations, Northeastern Montana." *Geology* **10:** 153–159. (p. 159)
12. Alvarez, W., L. W. Alvarez, et al. (1984). "The End of the Cretaceous: Sharp Boundary or Gradual Transition?" *Science* **223:** 1183–1186. (pp. 1183–1184)
13. Officer, C. B., and C. L. Drake (1983). "The Cretaceous–Tertiary Transition." *Science* **219:** 1383–1390. (p. 1385)
14. Ibid. (p. 1383)
15. Alvarez, W., L. W. Alvarez, et al. (1984). "The End of the Cretaceous: Sharp Boundary or Gradual Transition?" *Science* **223:** 1183–1186. (p. 1186)
16. Wezel, F. C., S. Vannucci, et al. (1981). "Découverte de divers niveaux riches en iridium dans 'la scaglia rossa' et 'la scaglia bianca' de l'Apennin d'Ombrie-Marches (Italie)." *Comptes-Rendus des séances de l'Académie des Sciences, série 2* **293:** 837–844.
17. Alvarez, W., L. W. Alvarez, et al. (1984). "The End of the Cretaceous: Sharp Boundary or Gradual Transition?" *Science* **223:** 1183–1186. (p. 1186)
18. Officer, C. B., and C. L. Drake (1985). "Terminal Cretaceous Environmental Events." *Science* **227:** 1161–1167. (p. 1161)
19. Smit, J., and F. Kyte (1985). "Cretaceous–Tertiary Extinctions: Alternative Models." *Science* **230:** 1292–1293. (p. 1292)

20. Crocket, J. H., C. B. Officer, et al. (1988). "Distribution of Noble Metals Across the Cretaceous/Tertiary Boundary at Gubbio, Italy: Iridium Variation as a Constraint on the Duration and Nature of Cretaceous/Tertiary Boundary Events." *Geology* **16:** 77–80.
21. Rocchia, R., D. Boclet, et al. (1990). "The Cretaceous–Tertiary Boundary at Gubbio Revisited: Vertical Extent of the Iridium Anomaly." *Earth and Planetary Science Letters* **99:** 206–219.
22. Alvarez, W., and F. Asaro (1990). "An Extraterrestrial Impact." *Scientific American* **263:** 78–84.
23. Ginsburg, R. N. (1995). Personal communication.
24. Anbar, A. D., G. J. Wasserburg, et al. (1996). "Iridium in Natural Waters." *Science* **273:** 1524–1528.
25. Kerr, R. A. (1987). "Asteroid Impact Gets More Support." *Science* **236:** 666–668. (p. 668)
26. Officer, C. B., and C. L. Drake (1985). "Terminal Cretaceous Environmental Events." *Science* **227:** 1161–1167. (p. 1164)
27. Izett, G. A. (1990). *The Cretaceous/Tertiary Boundary Interval, Raton Basin, Colorado and New Mexico, and Its Content of Shock-Metamorphosed Minerals.* Boulder, Colo.: U.S. Geological Survey.
28. Ibid. (p. 1164)
29. Ibid. (p. 1164)
30. French, B. M., and N. W. Short, Eds. (1968). *Shock Metamorphism of Natural Materials.* Baltimore: Mono Book Corporation.
31. French, B. M. (1985). "Cretaceous–Tertiary Extinctions: Alternative Models." *Science* **230:** 1293–1294. (p. 1293)
32. Carter, N. L., C. B. Officer, et al. (1986). "Dynamic Deformation of Volcanic Ejecta from the Toba Caldera: Possible Relevance to Cretaceous/Tertiary Boundary Phenomena." *Geology* **14:** 380–383.
33. Bohor, B., P. J. Modreski, et al. (1987). "Shocked Quartz in the Cretaceous–Tertiary Boundary Clays: Evidence for a Global Distribution." *Science* **236:** 705–709.
34. Alexopoulos, J. S., R. A. F. Grieve, et al. (1988). "Microscopic Lamellar Deformation Features in Quartz: Discriminative Characteristics of Shock-Generated Varieties." *Geology* **16:** 796–799.
35. Kerr, R. A. (1987). "Asteroid Impact Gets More Support." *Science* **236:** 666–668. (p. 667)
36. Ibid. (p. 667)
37. Carter, N. L., and C. B. Officer (1989). "Comment and Reply on Microscopic Lamellar Deformation Features in Quartz: Discriminative Characteristics of Shock-Generated Varieties." *Geology* **17:** 477–478. (p. 477)
38. Kerr, R. A. (1987). "Asteroid Impact Gets More Support." *Science* **236:** 666–668. (p. 666)
39. Officer, C. B., and N. L. Carter (1991). "A Review of the Structure, Petrology, and Dynamic Deformation Characteristics of Some Enigmatic Terrestrial Structures." *Earth-Science Reviews* **30:** 1–49. (p. 1)
40. Ibid. (p. 18)

41. Alexopoulos, J. S., R. A. F. Grieve, et al. (1988). "Microscopic Lamellar Deformation Features in Quartz: Discriminative Characteristics of Shock-Generated Varieties." *Geology* **16:** 796–799. (abstract, pp. 6–7)
42. Smit, J., and G. Klaver (1981). "Sanidine Spherules at the Cretaceous–Tertiary Boundary Indicate a Large Impact Event." *Nature* **292:** 47–49.
43. Montanari, A., R. L. Hay, et al. (1983). "Spheroids at the Cretaceous–Tertiary Boundary Are Altered Impact Droplets of Basaltic Composition." *Geology* **11:** 668–671.
44. Wezel, F. C., S. Vannucci, et al. (1981). "Découverte de divers niveaux riches en iridium dans 'la scaglia rossa' et 'la scaglia bianca' de l'Apennin d'Ombrie-Marches (Italie)." *Comptes-Rendus des séances de l'Académie des Sciences, série 2* **293:** 837–844.
45. Naslund. H. R., C. B. Officer, et al. (1986). "Microspherules in Upper Cretaceous and Lower Tertiary Clay Layers at Gubbio, Italy." *Geology* **14:** 923–926.
46. Browne, M. W. (January 19, 1988). "The Debate over Dinosaur Extinction Takes an Unusually Rancorous Turn." *New York Times,* pp. C1–C4. (p. C4)
47. Robin, E., D. Boclet, et al. (1991). "The Stratigraphic Distribution of Ni-Rich Spinels in Cretaceous–Tertiary Boundary Rocks at El-Kef (Tunisia), Caravaca (Spain) and Hole-761C (Leg-122)." *Earth and Planetary Science Letters* **107:** 715–721.
48. Glen, W. (1994). *The Mass Extinction Debates: How Science Works in a Crisis.* Stanford, Calif.: Stanford University Press. (p. 77)

CHAPTER 6

1. Orczy, E., Baroness (1974). *The Scarlet Pimpernel.* New York: Penguin Books. (p. 93)
2. Vogt, P. R. (1972). "Evidence for Global Synchronism in Mantle Plume Convection, and Possible Significance for Geology." *Nature* **240:** 338–342.
3. McLean, D. M. (1978). "A Terminal Mesozoic 'Greenhouse'; Lessons from the Past." *Science* **201:** 401–406.
4. Officer, C. B., and C. L. Drake (1985). "Terminal Cretaceous Environmental Events." *Science* **227:** 1161–1167.
5. Zoller, W. H., J. R. Parrington, et al. (1983). "Iridium Enrichment in Airborne Particles from Kilauea Volcano: January 1983." *Science* **222:** 1118–1120.
6. Officer, C. B., and C. L. Drake (1985). "Terminal Cretaceous Environmental Events." *Science* **227:** 1161–1167. (p. 1163)
7. Toutain, J.-P., and G. Meyer (1989). "Iridium-bearing Sublimates at a Hot-Spot Volcano (Piton de la Fournaise, Indian Ocean)." *Geophysical Research Letters* **16**(12): 1391–1394.
8. Felitsyn, S. B., and P. A. Vaganov (1988). "Iridium in the Ash of Kamchatkan Volcanoes." *International Geology Review* **30:** 1288–1291.

9. Koeberl, C. (1989). "Iridium Enrichment in Volcanic Dust from Blue Ice Fields, Antarctica, and Possible Relevance to the K/T Boundary Event." *Earth and Planetary Science Letters* **92:** 317–322.
10. Schmitz, B., and F. Asaro (1996). "Iridium Geochemistry of Ash Layers from Eocene Rifting." *Bulletin of the Geological Society of America* **108:** 489–504.
11. Rampino, M. R., and R. B. Stothers (1988). "Flood Basalt Volcanism During the Past 250 Million Years." *Science* **1988:** 663–668.
12. Courtillot, V. E., G. Feraud, et al. (1988). "Deccan Flood Basalts and the Cretaceous/Tertiary Boundary." *Nature* **333:** 843–846.
13. Bhandari, N., P. N. Shukla, et al. (1995). "Impact Did Not Trigger Deccan Volcanism: Evidence from Anjar K/T Boundary Intertrappean Sediments." *Geophysical Research Letters* **22:** 433–436.
14. McLean, D. (1993). "Asteroid or Volcano: Have the Volcanists Been Heard?" *Science* **259:** 877. (p. 877)
15. Koshland, D. E., Jr. (1993). "Asteroid or Volcano: Have the Volcanists Been Heard?" *Science* **259:** 877. (p. 877)
16. Alvarez, W. (1997). *T-Rex and the Crater of Doom.* Princeton: Princeton University Press. (p. 85)
17. McLean web page at http://www.vt.edu:10021/artsci/geology/mclean/Dinosaur_Volcano_Extinction/index.html.
18. Ibid.
19. Browne, M. W. (January 19, 1988). "The Debate over Dinosaur Extinction Takes an Unusually Rancorous Turn." *New York Times,* pp. C1–C4. (p. C4)

CHAPTER 7

1. Doyle, A. C. (1950). *Adventures of Sherlock Holmes,* Vol. 1. New York, Heritage. (p. 652)
2. French, B. M. (1984). "Impact Event at the Cretaceous–Tertiary Boundary: A Possible Site." *Science* **226:** 353.
3. Hildebrand, A. R. (1993). "The Cretaceous–Tertiary Boundary Impact (or the Dinosaurs Didn't Have a Chance)." *Journal of the Royal Astronomical Society of Canada* **87:** 77–118.
4. Koeberl, C., and R. R. Anderson (1996). Manson and Company: Impact Structures in the United States. *The Manson Impact Structure, Iowa: Anatomy of an Impact Crater,* eds. C. Koeberl and R. R. Anderson. Boulder, Colo.: Geological Society of America. (**Special Paper 302:** 1–29)
5. Kunk, M. J., G. A. Izett, et al. (1989). "Ar-40/Ar-39 Dating of the Manson Impact Structure: A Cretaceous–Tertiary Boundary Crater Candidate." *Science* **244:** 1565–1568.
6. Izett, G. A., W. A. Cobban, et al. (1993). "The Manson Impact Structure: Ar-40/Ar-39 Age and Its Distal Impact Ejecta in the Pierre Shale in Southeastern North Dakota." *Science* **262:** 729–732.
7. Ibid.

8. Hildebrand, A. R., and W. V. Boynton (1991). "Cretaceous Ground Zero." *Natural History*(6), 47–53. Hildebrand, A. R. (1993). "The Cretaceous–Tertiary Boundary Impact (or the Dinosaurs Didn't Have a Chance)." *Journal of the Royal Astronomical Society of Canada* **87:** 77–118.
9. Maurrasse, F. J.-M, and G. Sen (1991). "Impacts, Tsunamis, and the Haitian Cretaceous–Tertiary Boundary Layer." *Science* **252:** 1690–1693.
10. Hildebrand, A. R., and W. V. Boynton (1990) "Proximal Cretaceous–Tertiary Boundary Impact Deposits in the Caribbean." *Science* **248:** 843–847.
11. Hildebrand, A. R., and W. V. Boynton (1991). "Cretaceous Ground Zero." *Natural History*(6), 47–53. (p. 52)
12. Penfield, G. T. (1991). "Pre-Alvarez Impact." *Natural History* **(6),** 4. (p. 4)
13. Verschuur, G. L. (1996). *Impact: The Threat of Comets and Asteroids.* New York: Oxford University Press.
14. Alvarez, W. (1997). *T-Rex and the Crater of Doom.* Princeton, NJ: Princeton University Press.
15. Byars, C. (December 13, 1981). "Mexican Site May Be Link to Dinosaur's Disappearance." *Houston Chronicle,* pp. 1, 18.
16. "Possible Yucatan Impact Basin." *Sky and Telescope* **63:** 249–250. (p. 249)
17. Alvarez, W. (1997). *T-Rex and the Crater of Doom.* Princeton, NJ: Princeton University Press.
18. Sharpton, V. L., L. E. Marin, et al. (1996). A Model of the Chicxulub Impact Basin Based on Evaluation of Geophysical Data, Well Logs, and Drill Core Samples. *The Cretaceous–Tertiary Event and Other Catastrophes in Earth History,* eds. G. Ryder, D. Fastovsky, and S. Gartner. Boulder, Colo.: Geological Society of America. (**Special Paper 307:** 55–74.)
19. Pope, K. O., K. H. Baines, et al. (1997). "Energy, Volatile Production, and Climatic Effects of the Chicxulub Cretaceous/Tertiary Impact." *Journal of Geophysical Research,* E (*Planets*) **102,** 21,645–21,664.
20. Pope, K. O., A. C. Ocampo, et al. (1991). "Mexican Site for K/T Impact Crater." *Nature* **351:** 105.
21. Officer, C. B., C. L. Drake, et al. (1992). "Cretaceous–Tertiary Events and the Caribbean Caper." *GSA Today* **2:** 69–70, 73–74.
22. Verschuur, G. L. (1996). *Impact: The Threat of Comets and Asteroids.* New York: Oxford University Press.
23. Benton, M. J., and C. T. S. Little (1994). "Impact in the Caribbean and Death of the Dinosaurs." *Geology Today* **13:** 222–227.
24. Izett, G. A., G. B. Dalrymple, et al. (1991). "Ar-40/Ar-39 Age of Cretaceous–Tertiary Boundary Tektites from Haiti." *Science* **252:** 1539–1542.
25. Hall, C. M., D. York, et al. (1991). "Laser Ar-40/Ar-39 Step-Heating Ages from Single Cretaceous–Tertiary Boundary Glass Spherules." *Eos* **72:** A531.

26. Swisher, C. C. I., J. M. Grajales-Nichimura, et al. (1992). "Coeval Ar-40/Ar-39 Ages of 65.0 Million Years Ago from the Chicxulub Crater Melt Rock and Cretaceous–Tertiary Boundary Tektites." *Science* **257:** 954–958.
27. Sharpton, V. L., G. B. Dalrymple, et al. (1992). "New Links Between the Chicxulub Impact Structure and the Cretaceous/Tertiary Boundary." *Nature* **359:** 819–821.
28. Sigurdsson, H., S. D'Hondt, et al. (1991). "Glass from the Cretaceous/Tertiary Boundary in Haiti." *Nature* **349:** 482–487.
29. Blum, J. D., C. P. Chamberlain, et al. (1993). "Isotopic Comparison of K/T Boundary Impact Glass with Melt Rock from Chicxulub and Manson Impact Structures." *Nature* **364:** 325–327.
30. Ocampo, A. C., K. O. Pope, et al. (1996). Ejecta Blanket Deposits of the Chicxulub Crater from Albion Island, Belize. *The Cretaceous–Tertiary Event and Other Catastrophes in Earth History,* eds. G. Ryder, D. Fastovsky, and S. Gartner. Boulder, Colo.: Geological Society of America. (**Special Paper 307:** 75–88.)
31. Smit, J., and A. J. T. Romein (1985). "A Sequence of Events Across the Cretaceous–Tertiary Boundary." *Earth and Planetary Science Letters* **74:** 155–170.
32. Bourgeois, J., T. A. Hansen, et al. (1988). "A Tsunami Deposit at the Cretaceous–Tertiary Boundary in Texas." *Science* **241:** 567–570.
33. Smit, J., A. Montanari, et al. (1992). "Tektite-Bearing, Deep-Water Clastic Unit at the Cretaceous–Tertiary Boundary in Northeastern Mexico." *Geology* **20:** 99–103. (p. 103)
34. Stinnesbeck, W., J. M. Barbarin, et al. (1993). "Deposition of Channel Deposits Near the Cretaceous–Tertiary Boundary in Northeastern Mexico: Catastrophic or Normal Sedimentary Deposits?" *Geology* **21:** 797–800.
35. Ibid.
36. Fischer, A. (1996). Personal communication.
37. Kerr, R. (1994). "Testing an Ancient Impact's Punch." *Science* **263:** 1371–1372. (p. 1372)
38. Keller, G. (1994). "K–T Boundary Issues." *Science* **264:** 641. (p. 641)
39. Kerr, R. (1994). "K–T Boundary Issues." *Science* **264:** 642. (p. 642)
40. Meyerhoff, A. A., J. B. Lyons, et al. (1994). "Chicxulub Structure: A Volcanic Sequence of Late Cretaceous Age." *Geology* **22**(1): 3–4.
41. Officer, C. B., and J. Page (1996). *The Great Dinosaur Extinction Controversy.* Reading, Mass.: Addison-Wesley.
42. Ibid. (pp. 144, 147)
43. Ibid. (p. 155)
44. Ibid. (p. 155)
45. Hildebrand, A. R., G. T. Penfield, et al. (1991). "Chicxulub Crater: A Possible Cretaceous/Tertiary Boundary Impact Crater on the Yucatan Peninsula, Mexico." *Geology* **19:** 867–871.

46. Kerr, R. (1994). "Testing an Ancient Impact's Punch." *Science* **263:** 1371–1372. (p. 1372)
47. Bohor, B., W. J. Betterton, et al. (1993). "Impact-Shocked Zircons: Discovery of Shock-Induced Textures Reflecting Increasing Degrees of Shock Metamorphism." *Earth and Planetary Science Letters* **119:** 419–424.
48. Krogh, T. E., S. Kamo, et al. (1993). "U-Pb Ages of Single Shocked Zircons Linking Distal K/T Ejecta to the Chicxulub Crater." *Nature* **366:** 731–734.
49. Ibid. (p. 731)
50. Kamo, S. L., and T. E. Krogh (1995). "Chicxulub Crater Source for Shocked Zircon Crystals from the Cretaceous–Tertiary Boundary Layer, Saskatchewan; Evidence from New U-Pb Data." *Geology* **23:** 281–284.

CHAPTER 8

1. Darwin, C. (1964 Facsimile Edition). *On the Origin of Species by Means of Natural Selection.* Cambridge, Mass.: Harvard University Press. (pp. 319–320)
2. Alvarez, L. W. (1983). "Experimental Evidence That an Asteroid Impact Led to the Extinction of Many Species 65 Million Years Ago." *Proceedings of the National Academy of Sciences* **80:** 627–642. (p. 632)
3. Darwin, C. (1964 Facsimile Edition). *On the Origin of Species by Means of Natural Selection.* Cambridge, Mass.: Harvard University Press. (pp. 316–317)
4. Ibid. (p. 159)
5. Ibid. (p. 317)
6. Raup, D. M. (1992). *Extinction: Bad Genes or Bad Luck?* New York: Norton.
7. Raup, D. M. (1994). "The Role of Extinction in Evolution." *Proceedings of the National Academy of Sciences of the United States of America* **91:** 6758–6763. (p. 6758)
8. Jablonski, D. (1986). Causes and Consequences of Mass Extinctions; a Comparative Approach. *Dynamics of Extinction,* ed. D. K. Elliott. New York: John Wiley, pp. 183–229.
9. Raup, D. M., and J. J. Sepkoski, Jr. (1984). "Periodicity of Extinctions in the Geologic Past." *Proceedings of the National Academy of Sciences of the United States of America* **81**(3): 801–805.
10. Raup, D. M. (1992). *Extinction: Bad Genes or Bad Luck?* New York: Norton.
11. Signor, P. W. I., and J. H. Lipps (1982). Sampling Bias, Gradual Extinction Patterns, and Catastrophes in the Fossil Record. *Geological Implications of Impacts of Large Asteroids and Comets on the Earth,* eds. L. T. Silver and P. H. Schultz. Boulder, Colo.: Geological Society of America. (**Special Paper 190:** 291–296.)
12. Ibid.
13. Ibid. (p. 292)

14. Russell, D. A. (1982). "The Mass Extinctions of the Late Mesozoic." *Scientific American* **246**(January): 58–65.
15. Alvarez, L. W. (1983). "Experimental Evidence that an Asteroid Impact Led to the Extinction of Many Species 65 Million Years Ago." *Proceedings of the National Academy of Sciences* **80:** 627–642.
16. Williams, M. E. (1994). "Catastrophic Versus Noncatastrophic Extinction of the Dinosaurs: Testing, Falsifiability, and the Burden of Proof." *Journal of Paleontology* **68:** 183–190.
17. Meldahl, K. H. (1990). "Sampling, Species Abundance, and the Stratigraphic Signature of Mass Extinction: A Test Using Holocene Tidal Flat Molluscs." *Geology* **18:** 890–893.
18. Ibid.
19. Signor, P. W. I., and J. H. Lipps (1982). Sampling Bias, Gradual Extinction Patterns, and Catastrophes in the Fossil Record. *Geological Implications of Impacts of Large Asteroids and Comets on the Earth,* eds. L. T. Silver and P. H. Schultz. Boulder, Colo.: Geological Society of America. (**Special Paper 190:** 291–296.) (p. 291)

CHAPTER 9

1. Kerr, R. A. (1991). "Dinosaurs and Friends Snuffed Out?" *Science* **251:** 160–162. (p. 162)
2. Carlisle, D. B. (1995). *Dinosaurs, Diamonds, and Things from Outer Space.* Stanford, Calif.: Stanford University Press. (pp. 43–44)
3. Miller, W. (1950). *A Canticle for Leibowitz.* Philadelphia: Lippincott.
4. Ward, P. D. (1994). *The End of Evolution.* New York: Bantam Books.
5. Ibid.
6. Ward, P. D. (1983). "The Extinction of the Ammonites." *Scientific American* **249** (October): 136–147. (p. 136)
7. Ibid. (p. 136)
8. Ibid. (p. 147)
9. Ward, P. D. (1992). *On Methuselah's Trail.* New York: W. H. Freeman. (p. 119)
10. Ward, P. D. (1990). The Cretaceous/Tertiary Extinctions in the Marine Realm; a 1990 Perspective. *Global Catastrophes in Earth History,* eds. V. L. Sharpton and P. D. Ward. Boulder, Colo: Geological Society of America. (**Special Paper 247:** 425–432.) (p. 427)
11. Ibid. (p. 428)
12. Ward, P. D. (1994). *The End of Evolution.* New York: Bantam Books. (p. 154)
13. Marshall, C. R., and P. D. Ward (1996). "Sudden and Gradual Molluscan Extinctions in the Latest Cretaceous of Western European Tethys." *Science* **274:** 1360–1363.
14. Clemens, W. A., J. D. Archibald, et al. (1981). "Out With a Whimper Not a Bang." *Paleobiology* 7(3): 293–298.
15. Ibid. (pp. 297–298)

16. Upchurch, G. R. J. (1989). Terrestrial Environmental Changes and Extinction Patterns at the Cretaceous–Tertiary Boundary, North America. *Mass Extinctions; Processes and Evidence*, ed. S. K. Donovan. New York: Columbia University Press, pp. 195–216.
17. Clemens, W. A., J. D. Archibald, et al. (1981). "Out With a Whimper Not a Bang." *Paleobiology* **7**(3): 293–298. (p. 297)
18. Hickey, L. J. (1981). "Land Plant Evidence Compatible with Gradual, Not Catastrophic, Change at the End of the Cretaceous." *Nature* **292:** 529–531. (p. 529)
19. Orth, C. J., J. S. Gilmore, et al. (1981). "An Iridium Abundance Anomaly at the Palynological Cretaceous–Tertiary Boundary in Northern New Mexico." *Science* **214:** 1341–1343.
20. Johnson, K. R., and L. J. Hickey (1990). Patterns of Megafloral Change Across the Cretaceous–Tertiary Boundary in the Northern Great Plains and Rocky Mountains. *Global Catastrophes in Earth History*, eds. V. L. Sharpton and P. D. Ward. Boulder, Colo.: Geological Society of America. (**Special Paper 247:** 433–444.)
21. Ibid. (p. 433)
22. Kerr, R. A. (1991). "Dinosaurs and Friends Snuffed Out?" *Science* **251:** 160–162. (p. 162)
23. Johnson, K. R. (1995). "One Really Bad Day." *Denver Museum of Natural History Quarterly* **Spring 1995**: 2–5. (p. 3)
24. Archibald, J. D. (1996). *Dinosaur Extinction and the End of an Era: What the Fossils Say.* New York: Columbia University Press. (p. 180)
25. Johnson, K. R. (1997). Personal communication.
26. Thierstein, H. (1982). Terminal Cretaceous Plankton Extinctions: A Critical Assessment. *Geological Implications of Impacts of Large Asteroids and Comets on the Earth*, eds. L. T. Silver and P. H. Schultz. Boulder, Colo.: Geological Society of America. (**Special Paper 190:** 385–399.)
27. Smit, J. (1982). Extinction and Evolution of Planktonic Foraminifera After a Major Impact at the Cretaceous/Tertiary Boundary. *Geological Implications of Impacts of Large Asteroids and Comets on the Earth*, eds. L. T. Silver and P. H. Schultz. Boulder, Colo.: Geological Society of America. (**Special Paper 190:** 329–352.)
28. Keller, G., and E. Barrera (1990). The Cretaceous/Tertiary Boundary Impact Hypothesis and the Paleontological Record. *Global Catastrophes in Earth History*, eds. V. L. Sharpton and P. D. Ward. Boulder, Colo.: Geological Society of America. (**Special Paper 247:** 563–575.) (pp. 563, 566)
29. Fischer, A. (1996). Personal communication.
30. Kerr, R. (1994). "Testing an Ancient Impact's Punch." *Science* **263:** 1371–1372. (p. 1371)
31. Ibid. (pp. 1371–1372)
32. Ibid. (p. 1372)
33. Keller, G. (1994). "K–T Boundary Issues." *Science* **264:** 641.
34. Ibid. (p. 641)

35. Kerr, R. (1994). "K–T Boundary Issues." *Science* **264:** 642. (p. 642)
36. Ginsburg, R. N. (1995). Personal communication.
37. Keller, G. (1993). "The Cretaceous–Tertiary Boundary Transition in the Antarctic Ocean and Its Global Implications." *Marine Micropaleontology* **21:** 1–45.
38. Huber, B. T., C. Liu, et al. (1994). "Comment on the Cretaceous–Tertiary Boundary Transition in the Antarctic Ocean and Its Global Implications." *Marine Micropaleontology* **24:** 91–99.
39. Huber, B. T. (1996). Personal communication.
40. Ibid. Keller, G., and N. MacLeod (1994). "Reply to Comment on the Cretaceous–Tertiary Boundary Transition in the Antarctic Ocean and Its Global Implications." *Marine Micropaleontology* **24:** 101–118.
41. Huber, B. T. (1996). Personal communication.
42. Keller, G., and N. MacLeod (July 12, 1993). Letter to Robert Adams.
43. Adams, R. (October 6, 1993). Letter in reply to Keller and MacLeod.
44. MacLeod, N., and G. Keller, eds. (1996). *Cretaceous–Tertiary Mass Extinctions: Biotic and Environmental Changes.* New York: Norton.
45. MacLeod, N., and G. Keller, (1994). "Comparative Biogeographic Analysis of Planktic Foraminiferal Survivorship Across the Cretaceous/Tertiary (K/T) Boundary." *Paleobiology* **20:** 143–177.
46. O'Brian, P. (1979). *Desolation Island.* New York: Stein and Day.
47. Huber, B. T. (1991). Maestrichtian Planktonic Foraminifer Biostratigraphy and the Cretaceous/Tertiary Boundary at Hole 738C (Kerguelen Plateau, Southern Indian Ocean). *Proceedings of the Ocean Drilling Program,* eds. J. Barron and B. Larsen. College Station, Texas: Ocean Drilling Program. (**Leg 119:** 451–465.) Huber, B. T. (1996). Evidence for Planktonic Foraminifer Reworking Versus Survivorship Across the Cretaceous–Tertiary Boundary at High Latitudes. *The Cretaceous–Tertiary Event and Other Catastrophes in Earth History,* eds. G. Ryder, D. Fastovsky, and S. Gartner. Boulder, Colo.: Geological Society of America. (**Special Paper 307:** 319–334.)
48. Ibid. (1996, p. 332)

CHAPTER 10

1. Browne, M. W. (January 19, 1988). "The Debate over Dinosaur Extinction Takes an Unusually Rancorous Turn." *New York Times,* C1–C4. (p. C3)
2. Benton, M. J. (1990). "Scientific Methodologies in Collision: The History of the Study of the Extinction of the Dinosaurs." *Evolutionary Biology* **24:** 371–400.
3. Schopf, T. J. M. (1982). Extinction of the Dinosaurs: A 1982 Understanding. *Geological Implications of Impacts of Large Asteroids and Comets on the Earth,* eds. L. T. Silver and P. H. Schultz. Boulder, Colo.: Geological Society of America. (**Special Paper 190:** 415–422.) (pp. 415, 421)
4. Archibald, J. D., and W. A. Clemens (1982). "Late Cretaceous Extinctions." *American Scientist* **70:** 377–385. (p. 384)

5. Alvarez, L. W. (1983). "Experimental Evidence That an Asteroid Impact Led to the Extinction of Many Species 65 Million Years Ago." *Proceedings of the National Academy of Sciences* **80:** 627–642.
6. Ibid. (p. 627)
7. Ibid. (p. 627)
8. Browne, M. W. (January 19, 1988). "The Debate over Dinosaur Extinction Takes an Unusually Rancorous Turn." *New York Times,* p. C1–C4.
9. Jastrow, R. (1983). "The Dinosaur Massacre: A Double-Barrelled Mystery." *Science Digest* **September 1983:** 51–53, 109. (p. 109)
10. Alvarez, W. (1991). "The Gentle Art of Scientific Trespassing." *GSA Today* **1:** 29–31, 34.
11. Alvarez, L. W. (1983). "Experimental Evidence That an Asteroid Impact Led to the Extinction of Many Species 65 Million Years Ago." *Proceedings of the National Academy of Sciences* **80:** 627–642. (p. 635)
12. Ibid. (p. 639)
13. Archibald, J. D., and W. A. Clemens (1982). "Late Cretaceous Extinctions." *American Scientist* **70:** 377–385.
14. Alvarez, L. W. (1983). "Experimental Evidence That an Asteroid Impact Led to the Extinction of Many Species 65 Million Years Ago." *Proceedings of the National Academy of Sciences* **80:** 627–642. (p. 641)
15. Hoffman, A., and M. Nitecki (1985). "Reception of the Asteroid Hypothesis of Terminal Cretaceous Extinctions." *Geology* **13:** 884–887.
16. Raup, D. M. (1986). *The Nemesis Affair.* New York: Norton.
17. Browne, M. W. (October 29, 1985). "Dinosaur Experts Resist Meteor Extinction Idea." *New York Times,* pp. C1–C3.
18. Ibid.
19. Ibid. (p. C3)
20. Ibid. (p. C3)
21. Ibid. (p. C3)
22. Browne, M. W. (January 19, 1988). "The Debate over Dinosaur Extinction Takes an Unusually Rancorous Turn." *New York Times,* pp. C1–C4. (p. C4)
23. Bernal, J. P. (1939). *The Social Function of Science.* Cambridge, Mass.: M.I.T. Press.
24. Browne, M. W. (January 19, 1988). "The Debate over Dinosaur Extinction Takes an Unusually Rancorous Turn." *New York Times,* pp. C1–C4. (p. C4)
25. Ibid. (p. C4)
26. Alvarez, L. W. (1987). *Adventures of a Physicist.* New York: Basic Books.
27. Rhodes, R. (1995). *Dark Sun: The Making of the Hydrogen Bomb.* New York: Simon & Schuster.
28. Browne, M. W. (January 19, 1988). "The Debate over Dinosaur Extinction Takes an Unusually Rancorous Turn." *New York Times,* pp. C1–C4. (p. C4)

29. Ibid. (p. C4)
30. Hecht, J. (1988). "Evolving Theories for Old Extinctions." *New Scientist* **120** (November 12, 1988): 28–30. (p. 28)
31. Benton, M. J. (1990). "Scientific Methodologies in Collision: The History of the Study of the Extinction of the Dinosaurs." *Evolutionary Biology*, **24:** 371–400.
32. Raup, D. M. (1994). "The Role of Extinction in Evolution." *Proceedings of the National Academy of Sciences of the United States of America* **91:** 6758–6763. (p. 6758)
33. Raup, D. M. (1982). Biogeographic Extinction: A Feasibility Test. *Geological Implications of Impacts of Large Asteroids and Comets on the Earth*, eds. L. T. Silver and P. H. Schultz. Boulder, Colo.: Geological Society of America. (**Special Paper 190:** 277–282.) (p. 277)
34. Raup, D. M. (1992). *Extinction: Bad Genes or Bad Luck?* New York: Norton.
35. Fastovsky, D. E., and D. B. Weishampel (1996). *The Evolution and Extinction of the Dinosaurs.* New York: Cambridge University Press.
36. Archibald, J. D., and L. Bryant (1990). Differential Cretaceous/Tertiary Extinctions of Nonmarine Vertebrates: Evidence from Northeastern Montana. *Global Catastrophes in Earth History*, eds. V. L. Sharpton and P. D. Ward. Boulder, Colo.: Geological Society of America. (**Special Paper 247:** 549–562.)
37. Fastovsky, D. E., and P. M. Sheehan (1994). *Habitat vs. Asteroid Fragmentation in Vertebrate Extinctions at the KT Boundary; the Good, the Bad, and the Untested.* New Developments Regarding the KT Event and Other Catastrophes in Earth History, Houston, Texas, Lunar and Planetary Institute, Houston, TX. (p. 36)
38. Ward, P. D. (1995). "After the Fall: Lessons and Directions from the K/T Debate." *Palaios* **10:** 530–538. (p. 532)
39. Russell, D. A. (1979). "The Enigma of the Extinction of the Dinosaurs." *Annual Review of Earth and Planetary Sciences* **7:** 163–182.
40. Dodson, P. (1990). "Counting Dinosaurs; How Many Kinds Were There?" *Proceedings of the National Academy of Sciences of the United States of America* **87:** 7608–7612.
41. Russell, D. A. (1996). The Significance of the Extinction of the Dinosaurs. *The Cretaceous–Tertiary Event and Other Catastrophes in Earth History*, eds. G. Ryder, D. Fastovsky, and S. Gartner. Boulder, Colo.: Geological Society of America. (**Special Paper 307:** 381–388.)
42. Schopf, T. J. M. (1982). Extinction of the Dinosaurs: A 1982 Understanding. *Geological Implications of Impacts of Large Asteroids and Comets on the Earth*, ed. L. T. Silver and P. H. Schultz. Boulder, Colo.: Geological Society of America. (**Special Paper 190:** 415–422.)
43. Sloan, R. E., J. K. Rigby, Jr., et al. (1986). "Gradual Dinosaur Extinction and Simultaneous Ungulate Radiation in the Hell Creek Formation." *Science* **232:** 629–633.

44. Rigby, J. K., Jr., K. R. Newman, et al. (1987). "Dinosaurs from the Paleocene Part of the Hell Creek Formation, McCone County, Montana." *Palaios* **2:** 296–302.
45. Smit, J., and S. Van Der Kaars (1984). "Terminal Cretaceous Extinctions in the Hell Creek Area, Montana: Compatible with Catastrophic Extinction." *Science* **223:** 1177–1179.
46. Ibid.
47. Fastovsky, D. E. (1987). "Paleoenvironments of the Vertebrate-Bearing Strata During the Cretaceous–Paleocene Transition, Eastern Montana and Western North Dakota." *Palaios* **2:** 282–295.
48. Rigby, J. K., Jr., J. C. Ely, et al. (1995). "Paleomagnetic Evidence for Cretaceous–Tertiary Boundary Unconformity." *Abstracts with Programs—Geological Society of America Annual Meeting* **27:** 406.
49. Archibald, J. D., and L. Bryant (1990). Differential Cretaceous/Tertiary Extinctions of Nonmarine Vertebrates: Evidence from Northeastern Montana. *Global Catastrophes in Earth History*, eds. V. L. Sharpton and P. D. Ward. Boulder, Colo.: Geological Society of America. (**Special Paper 247:** 549–562.)
50. Sheehan, P. M., and D. E. Fastovsky (1992). "Major Extinctions of Land-Dwelling Vertebrates at the Cretaceous–Tertiary Boundary, Eastern Montana." *Geology* **20:** 556–560.
51. Sheehan, P. M., D. E. Fastovsky, et al. (1991). "Sudden Extinction of the Dinosaurs: Latest Cretaceous, Upper Great Plains, U.S.A." *Science* **254:** 835–839. (pp. 837–838)
52. Dodson, P., and L. P. Tatarinov (1990). Dinosaur Paleobiology; Part III, Dinosaur Extinction. *The Dinosauria*, eds. D. B. Weishampel, P. Dodson, and H. Osmólska. Berkeley, Calif.: University of California Press, pp. 55–62.
53. Clemens, W. A. (1992). "Dinosaur Extinction and Diversity." *Science* **256:** 159. (p. 159)
54. Dodson, P. (1990). "Counting Dinosaurs; How Many Kinds Were There?" *Proceedings of the National Academy of Sciences of the United States of America* **87:** 7608–7612. (p. 7612)
55. Glen, W. (1994). *The Mass Extinction Debates: How Science Works in a Crisis.* Stanford, Calif.: Stanford University Press. (p. 243)
56. Archibald, J. D. (1996). *Dinosaur Extinction and the End of an Era: What the Fossils Say.* New York: Columbia University Press.
57. Russell, D. A. (1996). "Dinosaur Extinction and the End of an Era (Book Review)." *Science* **273:** 1807–1808. (p. 1807)
58. Lockley, M. G., and A. P. Hunt (1996). *Dinosaur Tracks and Other Fossil Footprints of the Western United States.* New York: Columbia University Press.
59. Russell, D. A. (1996). "Dinosaur Extinction and the End of an Era (Book Review)." *Science* **273:** 1807–1808.
60. Ashraf, A. R., J. Stets, et al. (1995). "The Cretaceous–Tertiary Boundary in the Nanxiong Basin (Southeastern China)." *Abstracts with Programs—Geological Society of America Annual Meeting* **27:** 406–407.

61. Bhandari, N., P. N. Shukla, et al. (1995). "Impact Did Not Trigger Deccan Volcanism: Evidence from Anjar K/T Boundary Intertrappean Sediments." *Geophysical Research Letters* **22:** 433–436.
62. Toon, O. B., K. Zahnle, et al. (1994). Environmental Perturbations Caused by Impacts. *Hazards Due to Comets and Asteroids,* ed. T. Gehrels. Tucson: University of Arizona Press, 791–826.
63. Melosh, H. J., N. M. Schneider, et al. (1990). "Ignition of Global Wildfires at the Cretaceous/Tertiary Boundary." *Nature* **343:** 251–254. (p. 253)
64. Pope, K. O., K. H. Baines, et al. (1997). "Energy, Volatile Production, and Climatic Effects of the Chicxulub Cretaceous/Tertiary Impact." *Journal of Geophysical Research (Planets)* **102:** 21, 645–21, 664.
65. Retallack, G. J. (1996). "Acid Trauma at the Cretaceous–Tertiary Boundary in Eastern Montana." *GSA Today* **6:** 1–7.
66. MacLeod, N. (1996). "K/T Redux." *Paleobiology* **22:** 311–317. (p. 315)
67. Officer, C. B., and J. Page (1996). *The Great Dinosaur Extinction Controversy.* Reading, Mass.: Addison-Wesley. (pp. 177–178)

CHAPTER 11

1. Alvarez, L. W. (1983). "Experimental Evidence That an Asteroid Impact Led to the Extinction of Many Species 65 Million Years Ago." *Proceedings of the National Academy of Sciences* **80:** 627–642. (p. 633)
2. Vickery, A., and H. J. Melosh (1996). Atmospheric Erosion and Impactor Retention in Large Impacts, with Application to Mass Extinctions. *The Cretaceous–Tertiary Event and Other Catastrophes in Earth History,* eds. G. Ryder, D. Fastovsky, and S. Gartner. Boulder, Colo.: Geological Society of America. (**Special Paper 307:** 289–300.)
3. McGhee, G. R. (1996). *The Late Devonian Mass Extinction.* New York: Columbia University Press.
4. Koeberl, C., R. A. Armstrong, et al. (1997). "Morokweng, South Africa: A Large Impact Structure of Jurassic–Cretaceous Age." *Geology* **25:** 731–734.
5. Monastersky, R. (1997). "Life's Closest Call." *Science News* **151:** 74–75. (p. 74)
6. Kerr, R. A. (1996). "A Shocking View of the Permo–Triassic." *Science* **274:** 1080.
7. Monastersky, R. (1997). "Life's Closest Call." *Science News* **151:** 74–75.
8. Erwin, D. H. (1994). "The Permo–Triassic Extinction." *Nature* **367:** 231–235.
9. Renne, P. R., Z. Zichao, et al. (1995). "Synchrony and Causal Relations Between Permian–Triassic Boundary Crises and Siberian Flood Volcanism." *Science* **269:** 1413–1416.
10. Erwin, D. H. (1996). "The Mother of Mass Extinctions." *Scientific American* **275:** 72–78.
11. Knoll, A. H., R. K. Bambach, et al. (1996). "Comparative Earth History and the Late Permian Mass Extinction." *Science* **273:** 452–457.

12. Fowell, S. J., and P. E. Olsen (1993). "Time Calibration of Triassic/Jurassic Microfloral Turnover, Eastern North America." *Tectonophysics* **222:** 361–369.
13. Bottomley, R., R. A. F. Grieve, et al. (1997). "The Age of the Popigai Impact Event and Its Relation to Events at the Eocene/Oligocene Boundary." *Nature* **388:** 365–368.
14. Poag, C. W., D. S. Powars, et al. (1994). "Meteoroid Mayhem in Old Virginny: Source of the North American Tektite Strewn Field." *Geology* **22:** 691–694.
15. Prothero, D. R. (1994). *The Eocene–Oligocene Transition: Paradise Lost.* New York: Columbia University Press.
16. Raup, D. M. (1992). *Extinction: Bad Genes or Bad Luck?* New York: Norton.
17. Sepkoski, J. J., Jr. (1982). *A Compendium of Fossil Marine Families.* Milwaukee: Milwaukee Public Museum.
18. Raup, D. M. (1992). *Extinction: Bad Genes or Bad Luck?* New York: Norton.
19. Ibid.
20. Ibid.
21. Rampino, M. R., and B. M. Haggerty (1994). Extraterrestrial Impacts and Mass Extinctions of Life. *Hazards Due to Comets and Asteroids,* ed. T. Gehrels. Tucson: University of Arizona Press, pp. 827–858.
22. Raup, D. M. (1992). *Extinction: Bad Genes or Bad Luck?* New York: Norton.
23. Eldredge, N. (1996). "What Drives Evolution?" *Earth* **5:** 34–37.
24. Eldredge, N., and S. J. Gould (1972). Punctuated Equilibria; An Alternative to Phyletic Gradualism. *Models in Paleobiology.* San Francisco: Freeman, Cooper, pp. 82–115.
25. Eldredge, N. (1996). "What Drives Evolution?" *Earth* **5:** 34–37. (p. 36)

CHAPTER 12

1. Sepkoski, J. J., Jr. (1994). What I Did with My Research Career: Or How Research on Biodiversity Yielded Data on Extinction. *The Mass Extinction Debates: How Science Works in a Crisis,* ed. W. Glen. Stanford, Calif.: Stanford University Press, pp. 132–144. (p. 143)
2. Ibid.
3. Boswell, J. (1992). *The Life of Samuel Johnson.* New York: Knopf. (p. 118)
4. Sepkoski, J. J., Jr. (1994). What I Did with My Research Career: Or How Research on Biodiversity Yielded Data on Extinction. *The Mass Extinction Debates: How Science Works in a Crisis,* ed. W. Glen. Stanford, Calif.: Stanford University Press, pp. 132–144. (p. 143)
5. Raup, D. M., and J. J. Sepkoski, Jr. (1984). "Periodicity of Extinctions in the Geologic Past." *Proceedings of the National Academy of Sciences of the United States of America* **81**(3): 801–805.
6. Fischer, A. G., and M. A. Arthur (1977). Secular Variations in the Pelagic Realm. *Deepwater Carbonate Environments: Based on a Symposium Sponsored by the Society of Economic Paleontologists and Mineralogists,* eds.

H. E. Cook and P. Enos. Tulsa, Okla.: Society of Economic Paleontologists and Mineralogists, pp. 19–50.

7. Raup, D. M., and J. J. Sepkoski, Jr. (1984). "Periodicity of Extinctions in the Geologic Past." *Proceedings of the National Academy of Sciences of the United States of America* **81**(3): 801–805.
8. Alvarez, W., and R. A. Muller (1984). "Evidence from Crater Ages for Periodic Impacts on the Earth." *Nature* **308:** 718–720. Davis, M., P. Hut, et al. (1984). "Extinction of Species by Periodic Comet Showers." *Nature* **308:** 715–717. Rampino, M. R., and R. B. Stothers (1984). "Terrestrial Mass Extinctions, Cometary Impacts and the Sun's Motion Perpendicular to the Galactic Plane." *Nature* **308:** 709–712. Schwartz, R. D., and P. B. James (1984). "Periodic Mass Extinctions and the Sun's Oscillation about the Galactic Plane." *Nature* **308:** 712–713. Whitmire, D. P., and A. A. I. Jackson (1984). "Are Periodic Mass Extinctions Driven by a Distant Solar Companion?" *Nature* **308:** 713–715.
9. Davis, M., P. Hut, et al. (1984). "Extinction of Species by Periodic Comet Showers." *Nature* **308:** 715–717.
10. Gould, S. J. (1984). "The Cosmic Dance of Siva." *Natural History* **93:** 14–19. (pp. 18–19)
11. Ibid. (p. 19)
12. Whitmire, D. P., and A. A. I. Jackson (1984). "Are Periodic Mass Extinctions Driven by a Distant Solar Companion?" *Nature* **308:** 713–715.
13. Hoffman, A. (1985). "Patterns of Family Extinction Depend on Definition and Geological Timescale." *Nature* **315:** 659–662.
14. Maddox, J. (1985). "Periodic Extinctions Undermined." *Nature* **315:** 627. (p. 627)
15. Gould, S. J. (1985). "All the News That's Fit to Print and Some Opinions That Aren't." *Discover* **November 1985:** 86–91.
16. Hoffman, A. (1985). "Patterns of Family Extinction Depend on Definition and Geological Timescale." *Nature* **315:** 659–662. (p. 661)
17. Erwin, D. H. (1994). "The Permo–Triassic Extinction." *Nature* **567:** 231–235.
18. Raup, D. M., and J. J. Sepkoski, Jr. (1984). "Periodicity of Extinctions in the Geologic Past." *Proceedings of the National Academy of Sciences of the United States of America* **81**(3): 801–805.
19. Rampino, M. R., and R. B. Stothers (1984). "Terrestrial Mass Extinctions, Cometary Impacts and the Sun's Motion Perpendicular to the Galactic Plane." *Nature* **308:** 709–712.
20. Stothers, R. B. (1993). "Impact Cratering at Geologic Stage Boundaries." *Geophysical Research Letters* **20:** 887–890.
21. Grieve, R. A. F. (1997). Personal communication.
22. Rampino, M. R., and B. M. Haggerty (1995). "The Shiva Hypothesis: Impacts, Mass Extinctions, and the Galaxy." *Earth, Moon, & Planets* **72:** 441–460. Rampino, M. R., and B. M. Haggerty (1996). Impact Crises and Mass Extinctions: A Working Hypothesis. *The Cretaceous–Tertiary Event and Other Catastrophes in Earth History*, eds. G. Ryder, D. Fastovsky, and

S. Gartner. Boulder, Colo.: Geological Society of America. (**Special Paper 307:** 11–38.)
23. Stothers, R. B., and M. R. Rampino (1990). Periodicity in Flood Basalts, Mass Extinctions, and Impacts; A Statistical View and a Model. *Global Catastrophes in Earth History,* eds. V. L. Sharpton and P. D. Ward. Boulder, Colo. Geological Society of America. (**Special Paper 247:** 9–18.)
24. Courtillot, V. E. (1994). "Mass Extinctions in the Last 300 Million Years: One Impact and Seven Flood Basalts?" *Israel Journal of Earth Science* **43:** 255–266.
25. Sepkoski, J. J., Jr. (1989). "Periodicity in Extinction and the Problem of Catastrophism in the History of Life." *Journal of the Geological Society of London* **146:** 7–19.
26. Matese, J. J., P. G. Whitmire, et al. (1995). "Periodic Modulation of the Oort Cloud Comet Flux by the Adiabatically Changing Galactic Tide." *Icarus* **116:** 255–268.
27. Szpir, M. (1997). "Perturbing the Oort Cloud." *American Scientist* **85:** 24–26.
28. Ibid. (p. 24)
29. Erwin, D. H. (1994). *The Great Paleozoic Crisis.* New York: Columbia University Press. (p. 272)

CHAPTER 13

1. Shoemaker, E. M. (1997). Why Study Impact Craters? *Impact and Explosion Cratering,* eds. D. Roddy, R. O. Pepin, and R. B. Merrill. New York: Pergamon, pp. 1–10. (p. 1)
2. Marvin, U. B. (1990). Impact and Its Revolutionary Implications for Geology. *Global Catastrophes in Earth History,* eds. V. L. Sharpton and P. D. Ward. Boulder, Colo.: Geological Society of America. (**Special Paper 247:** 147–154.) (p. 147)
3. Jablonski, D. (1986). Causes and Consequences of Mass Extinctions; a Comparative Approach. *Dynamics of Extinction,* ed. D. K. Elliott. New York: John Wiley, pp. 183–229.
4. Kerr, R. A. (1997). "Climate-Evolution Link Weakens." *Science* **276:** 1968.
5. Kuhn, T. S. (1970). *The Structure of Scientific Revolutions.* Chicago: University of Chicago Press. (p. 90)
6. Officer, C. B., and J. Page (1996). *The Great Dinosaur Extinction Controversy.* Reading, Mass.: Addison-Wesley.
7. Ibid. (p. viii)
8. Ibid. (p. 157)
9. Ibid. (p. 60)
10. Ibid. (pp. 174–177)
11. Ibid. (p. 97)
12. Ibid. (p. 97)
13. Ibid. (p. 185)
14. Ward, P. D. (1995). "After the Fall: Lessons and Directions from the K/T Debate." *Palaios* **10:** 530–538.
15. Horgan, J. (1996). *The End of Science.* Reading, Mass.: Addison-Wesley.

INDEX